Science, Technology and the Ageing Society

Ageing is widely recognised as one of the social and economic challenges in the contemporary, globalised world, for which scientific, technological and medical solutions are continuously sought. This book proposes that science and technology also played a crucial role in the creation and transformation of the ageing society itself.

Drawing on existing work on science, technology and ageing in sociology, anthropology, history of science, geography and social gerontology, *Science, Technology and the Ageing Society* explores the complex, interweaving relationship between expertise, scientific and technological standards and social, normatively embedded age identities. Through a series of case studies focusing on older people, science and technology, medical research about ageing and ageing-related illnesses, and the role of expertise in the management of ageing populations, Moreira challenges the idea that ageing is a problem for the individual and society.

Tracing the epistemic and technological infrastructures that underpin multiple of ways of ageing, this timely volume is a crucial tool for undergraduate and graduate students interested in social gerontology, health and social care, sociology of ageing, science and technology studies and medical sociology.

Tiago Moreira is a Reader in Sociology at Durham University, UK.

Routledge Advances in Sociology

For a full list of titles in this series, please visit www.routledge.com/series/SE0511

Science, Technology and the Ageing Society

Tiago Moreira

Routledge
Taylor & Francis Group

LONDON AND NEW YORK

First published 2017 by Routledge

2 Park Square, Milton Park, Abingdon, Oxfordshire OX14 4RN
711 Third Avenue, New York, NY 10017

Routledge is an imprint of the Taylor & Francis Group, an informa business

First issued in paperback 2018

British Library Cataloguing in Publication Data
A catalogue record for this book is available from the British Library

Library of Congress Cataloging in Publication Data
Names: Moreira, Tiago, author.
Title: Science, technology and the ageing society / Tiago Moreira.
Description: Abingdon, Oxon; New York, NY: Routledge, 2017. |
Series: Routledge advances in sociology; 201 | Includes
bibliographical references and index.
Identifiers: LCCN 2016029987 | ISBN 9781138814127 (hardback)
Subjects: LCSH: Population aging–Social aspects. |
Gerontechnology. | Science–Social aspects. | Technology–Social
aspects.
Classification: LCC HQ1061.M5876 2017 | DDC 306–dc23
LC record available at https://lccn.loc.gov/2016029987

ISBN: 978-1-138-81412-7 (hbk)
ISBN: 978-1-318-34466-2 (pbk)

Typeset in Times New Roman
by Wearset Ltd, Boldon, Tyne and Wear

Edad, edad, tus venenosos liquidos.
Edad, edad, tus animales blancos.[1]

Antonio Gamoneda

---

1   Age, age, your poisonous liquids. / Age, age, your white animals.

Contents

Acknowledgements

The idea for this book was developed during a sabbatical research period hosted by the Centre for Healthy Ageing (CEHA), Copenhagen University in 2012, particularly during long walks and short bicycle rides between Christianshavn, Amagerbro, Amager Strand and Islands Brygge. I thank NORDEA Fonden for generously supporting my sabbatical, and my collaboration with CEHA since 2012. I am grateful to the members of CEHA's Health Innovation and Promotion Research Theme for their enthusiasm and critical engagement with my ideas and research, and in particular to Michael Andersen, Amy Clotworthy, Astrid Pernille Jespersen, Aske Juul Lassen, Lene Otto, Marie Haulund Otto and Thomas Söderqvist. I also thank Lene Juel Rasmussen, CEHA's Director, for enabling and fostering interdisciplinary collaboration in CEHA, and for including me in this.

As will be clear in the book itself, this book's outlook is inspired by the work of John Law and Annemarie Mol. Having had the privilege of working with both, their impact on my work has been more than conceptual or methodological. John was a generous and supportive doctoral supervisor and has continued to offer intellectual guidance and advice since then. He has been a moral and professional example. Annemarie was my 'unofficial' doctoral supervisor and has continued to challenge my ways of thinking and writing until today. I am a better thinker and writer for it.

I owe my introduction to the sciences of ageing to Tom Kirkwood. Through his singular vision and knowledge, Tom led directly to my interest in the role of science and technology in the 'ageing society'. In the long generation and development of this book, Tom has been at times a collaborator, an informant, a critical reader and a sponsor.

Many of the ideas presented in this book were developed with or alongside Paolo Palladino. He has consistently pushed me to question the way the 'past' and the 'future' and 'truth' and 'hope' figure in the stories we tell about the role of science and technology in the 'ageing society'. At times our continuing conversations about ageing and death have been the pretext for more important discussions on landscapes, food, animals, or walking paths, but that is the way it should be.

The book has also benefitted from discussions – however small – with a variety of people about ageing: John Bond, Katie Brittain, Alberto Cambrosio,

Michel Callon, Annie L. Cot, Tomas Sanchez Criado, Miguel Domènech, Jean Paul Gaudillière, Daniel Lopez Gomez, Claes-Friedrich Helgesson, Sam Hillyard, Julian Hughes, Stephen Katz, Joanna Latimer, Francis Lee, Thomas C. Leonard, Carl May, Robert McCrae, Ian McKeith, Richard Miller, Richard Milne, Maggie Mort, Ingunn Moser, João Passos, Martyn Pickersgill, Jeanette Pols, Vololona Rabeharisoa, Tim Rapley, Suresh Rattan, Jean-Marie Robine, Robin Wolfe Scheffler, Michael Schillmeier, Steve Sturdy, Bryan Turner, Kate Weiner, Rudi Westendorp, Catherine Will, Duncan Wilson, Andrew Webster, Miriam Winance and Teun Zuiderent-Jerak.

Particular chapters or parts of the book have gained from the insightful reading and comments of a few people: Geoff Bowker, Lawrence Busch, Michaela Fay, Neil Jenkings, Fadhila Manzaderani, Barbara Marshal, Annemarie Mol, Paolo Palladino, Hyung Wook Park, Richard Sprott and John Vincent. Arguments and sections of the book were presented in a variety of academic forums: British Gerontology Society Annual Conference 2015; British Sociological Association Annual Conference 2013; Centre for Medical Science and Technology Studies Copenhagen University; Durham University Department of Anthropology; Edinburgh University School of Social and Political Sciences; Kent University School of Sociology and Social Policy; London School of Hygiene and Tropical Medicine; Manchester University Centre for the History of Science, Technology and Medicine; Centre de Recherche Médicale et Sanitaire at Paris University; Sheffield University Department of Sociological Studies; Society for the Social Study of Science Annual Conference 2012 and 2014.

I could not have written the book without the support of Michaela Fay. Michaela patiently listened to ideas and extracts of text, and provided academic feedback and guidance on key aspects of the book. She also made sure that I balanced writing with my other pleasures and responsibilities, including that of being her partner and co-parent. Our daughters – Filipa and Marta – tracked the progress of the book with enthusiasm and encouragement. They have also continued to make my own process of getting older much richer and enjoyable. At a distance, but no less important, was the support of my mother, father, brothers and grandmother.

Versions of the material explored in the book have been published as journal papers, including:

*   Moreira, T. (2016) '(2016) 'De-standardising ageing? Shifting regimes of age measurement', *Ageing and Society*, 36(7), pp. 1407–1433
*   Moreira, T. (2015) 'Unsettling Standards: The biological age controversy', *The Sociological Quarterly*, 56(1): 18–39.
*   Moreira, T. and Palladino, P. (2011) ' "Population Laboratories" or "Laboratory Populations"? Making Sense of the Baltimore Longitudinal Study of Aging, 1965–1987', *Studies in History and Philosophy of Science Part C*, 42(3): 317–327.

Chapter 1

Science, technology and the 'ageing society'

This book focuses on the knowledge tools and technological means used to understand and manage ageing in contemporary societies. It takes as a point of departure the idea that while science and technology are mainly perceived as solutions to the 'problems of ageing' they have also contributed significantly to defining and articulating ageing as a societal issue, that is to say, to shaping the institutions and practices of what we usually label the 'ageing society'. The book investigates this interactive, co-productive relationship between science, technology and the 'ageing society'. As a consequence of this approach, it proposes that the concept of the 'ageing society' should refer not only to the socio-economic transformation associated with demographic ageing but also to the shifting relations between knowledge making practices, innovation processes and life course institutions. The 'ageing society', I argue, is first and foremost a collective predicament, a swelling uncertainty concerning how to deploy procedures of scientific research and technological innovation in addressing ageing as an issue. Ageing unsettles the scientific and technological apparatus of contemporary society. In contrast to other sociological approaches to this issue, which emphasise the increased encroachment of science, technology and medicine on the life course, this book stresses the controversial, unstable character of those ways of knowing and managing ageing. It does so by exploring the composition, the difficult coming together, of particular features of the epistemic and technological infrastructure of the 'ageing society': the disputes within population science on how to explain and address demographic ageing, the convoluted and still unsuccessful search for new age standards to measure individuals' evolving abilities and needs, the negotiated attempts to develop new research methods to measure individuals' unique ageing trajectory, the troubles associated with developing tools to support longer working and active lives, the unsteady character of the relationship between the care of older people and technologies aiming to support it, the critical challenges posed by ageing science to the social, economic and political organisation of contemporary biomedicine and health care, and the related moral and political divergences arising from proposals to extend human longevity. Taken together, these provide a conceptually robust analysis of how experts, policy makers, commercial and civil society organisations and citizens bring ageing to bear on the normative and material contexts of contemporary science and technology.

What is the 'ageing society'?

Public, policy and media discourses usually identify ageing and the 'ageing society' as one of the main societal challenges faced by contemporary societies. In a fairly typical example of this phenomenon, in March 2013 the BBC reported that the House of Lords Committee on 'public service and demographic change' had warned that the United Kingdom was 'woefully unprepared' for an ageing society, warning that the 'gift of a longer life' could lead to 'a series of crises' in public service provision (BBC, 2013). Drawing on figures from the Office of National Statistics and statements from experts, politicians and charities, it was reported that the Lords Committee had identified a rising trend in the number and proportion of older citizens, and called on political parties to explicitly address how such a trend could challenge social security schemes and health and social care provision. Summarising this call for action, Baroness Greengross, then chief executive of the International Longevity Centre UK – a think-tank focused on ageing and demographic change – stated that, as a society, 'we have put off the difficult decisions for far too long' and that it was now policy makers' responsibility to make the changes necessary in the welfare system to ameliorate 'the problems'.

There are three key dimensions that make this event typical of public discourse on the 'ageing society'. One is the factual nature of the demographic change that is being reported: demographic ageing is a reality on which experts, civic organisations and politicians of different inclinations agree. The second relates to how this change entails consequences for public institutions and systems that are at the heart of modern, advanced democratic societies: retirement age, old age pensions, health and social care provision, etc. Finally, there is the urgency with which such calls for action on the 'ageing society' are pronounced: 'there is no time to lose', no space for further debates or deliberation. It is now time to tackle the 'ageing society'. These three dimensions work together to make the 'ageing society' a much talked about but seldom examined concept or idea. What is the 'ageing society'? Why does it entail dramatic changes to the way in which work, family life and social and welfare institutions are organised?

One possible answer to the first of these questions would be to provide a widely used definition of the 'ageing society' – often attributed to the United Nations (2001) – which describes the category on the basis of a percentage of the total population in any given country (7 per cent) being 65 or more years of age. This is seen to indicate a threshold in a population's sustainability, that is to say, to specify the levels of fertility and longevity affecting a population's ability to maintain its current volume. Underpinning this assessment of population dynamics lays the claim that a decline in the average number of children born per woman, combined with an increase in life expectancy – as observed in most developed economies – is associated with a growth in the number and proportion of older people. This, in turn, furthers the decline in fertility levels, as reproduction is mostly expected to occur below the ages of 30–40.

This dynamic state is furthermore generally seen to have consequences for the social and economic relationships between age groups within a society. Because economic participation declines in later life, economic vitality is expected to deteriorate, leading to rising 'dependency ratios' – the proportion of money that has to be transferred from active to inactive sections of the population. From this widely accepted perspective, in such conditions active workers will have to increase productivity and to generate more income to sustain older, 'dependent' non-workers through pension schemes and other welfare programmes. Additionally, because of the increased burden of illness associated with later life, population ageing is also regarded as having straining effects on demand for health care services and funding of health insurance schemes. In the same way, declines in functional capacities associated with 'old age' are linked to increased pressures on systems of formal care and on processes of informal care within families and communities. Finally, ageing populations are said to produce consequences in the political and cultural attitudes of a society, which becomes more conservative and less open to social and technological innovation.

While most would agree with the factual assessment regarding the demographic forces and magnitude of population ageing, there is, however, a growing academic and policy debate about its social, economic and political consequences. Denouncing what Robertson (1990) once labelled 'apocalyptic demography', scholars and policy analysts have pointed to the problematic way in which ageing populations have become linked to emerging problems in public finances, and have proposed alternative scenarios of social and economic change. Harper (2006), a leading social gerontologist, suggested most social implications of demographic ageing are based on flawed or rigid assumptions about older people and the ageing process. She argues, for example, that inflexible retirement systems are more to blame for rising dependency ratios than older people's unwillingness or incapacity to work. Equally, Harper proposes that health care systems' rising costs are as much explained by ageing populations as by changes in lifestyle, public expectations in relation to medicine and increased use of health technologies. Finally, she proposes that, if anything, there is continuing commitment of younger generations to the care of older relatives or friends – albeit differentially distributed across genders – and that there is little evidence of decreasing intergenerational solidarity. In alignment with this view, in the much-debated book *The Imaginary Time Bomb*, Mullan (2000) argues that, from a historical perspective, population ageing has been a consistent feature of modern societies, being observable in countries such as France or Sweden at the turn of the twentieth century. Questioning the assumptions of the economic models where the present 'ageing society crisis' is based, Mullan denounces the erroneous assumptions on employment, productivity, economic dependency of older people, or the health and functional status of older people implied in those models. These scenarios, he argues, are based on the assumption that old age leads to functional decline, whereas contemporary population statistics point towards a more plastic relationship between ageing and health or disability (see Chapter 3).

Based on the same, optimistic understanding of the relationship between ageing and functional ability, demographers, epidemiologists and policy analysts have, for at least two decades, emphasised the opportunities as well as the challenges arising from population ageing. They highlight the benefits of moving away from age-segregated societies, where roles and capacities are ascribed in terms of belonging to an age group ('the elderly'), towards increasingly age-integrated social relations. These opportunities are linked to flexible retirement policies, the promotion of active involvement of older citizens in economic and social life, intergenerational assimilation, and support for health maintenance activities and social relations, aims encapsulated in the 'active ageing' policy frameworks of the European Union and the World Health Organization (Kalache and Kickbusch, 1997; Fernández-Ballesteros et al., 2013; Lassen and Moreira, 2014). Active and healthy ageing policies wish to articulate strategies for optimising economic, social and cultural participation throughout the life course, emphasising the societal gains yielded through 'maturing' social relations, such as political and cultural stability, allowing for normative and worldview connections between generations.

In addition, some expert proposals go as far as suggesting that increases in human longevity accrue societal bonuses – the so-called 'longevity dividend' (Olshansky et al., 2007). They argue that by supporting healthy behaviours and lifestyles, instruments and policies of active ageing can extend working and consuming lives, sustaining so-called mature or 'silver markets'. Health becomes thus an economic investment (Olshansky et al., 2012). Crucially, Olshansky and colleagues suggest that, in knowledge-based economies, age-integrated workplaces would lead to an efficient use of the accumulated experience and wisdom on the steering of innovation. Last but not least, older people's time and wisdom can be used to embolden and cement intergenerational relationships. The 'longevity dividend' proposal specifically aims to draw on the 'science of ageing' to achieve or extend a deceleration in the rate of ageing, so as to delay all ageing-related diseases and disorders by about seven years and halve the age-specific incidence of illness in later life. Bringing together biology, epidemiology, social demography, economics and policy, the 'longevity dividend' idea crosses disciplinary boundaries to outline a radical societal horizon where increases in longevity are associated with economic sustainability and social renewal.

The contrast between the pessimistic and the optimistic versions of the ageing society could not be more striking and is indicative of the significance of ageing for contemporary societies. It is indicative of this because both sides seem to agree that our collective future hinges, in a fundamental way, upon ageing. To what extent is ageing the key for our impending social, cultural and economic ways of living? To answer this question, we need to examine further the means through which claims about the opportunities and/or challenges of the ageing society are made. One approach to this would be to argue that either – or both – positions on the ageing society have got it somehow wrong, that their data is in fact incorrect, that their analysis is flawed, their assumptions biased, or influenced by vested interests. There are a variety of books, including Harper's (2006), which focus on this type of critical evaluation of claims about the 'ageing society'. This book aims to do something different. It asks the question: what if the controversies about the 'ageing society' were not only a sign of the importance of ageing in our collective life but also shaped the way in which we approach, understand and manage ageing processes in society?

Contrary to what is usually assumed, controversies – particularly public disputes of a technical variety hinging on a matter that concerns most people in society – are not mere surface phenomena, denoting only the biases of the opposing sides. Instead, controversies can be seen as key windows to understand how the society we live in is put together, or how, in this process of making the social, knowledge, technology and society become interlinked. As Latour put it, controversies are 'not simply a nuisance to be kept at bay, but what allows the social to be established' (Latour, 2005: 25). This is because, among other things, they substantiate ways of knowing and acting on the world, and mould institutions, policies and programmes. From a social science perspective, they give us direct insight into institution building, group formation, or identity construction processes (Latour, 2005; Venturini, 2010). That is to say, they give us a way to trace and explore how the 'ageing society' was put together. My proposal, throughout this book, is that this insight only comes to fruition if we delve deep into the *knowledge making practices, tools, technologies, the knowledge making institutions* that underpin positions within controversies about the 'ageing society'. This proposal veers however from the usual understanding of the role of science and technology in the 'ageing society'.

Typically, science and technology are viewed mainly as solutions to the 'problems' or challenges of the ageing society, discussed above. Hence, in the past two decades or so, ageing research has become increasingly acknowledged as a key area to focus economic and intellectual resources. This focus is seen as essential to generate the technological, medical and social innovations to address the age-related processes that structure public finances, health care systems and social care programmes, and seemingly affect the experience of 'old age' for a large proportion of the population. New fields of research, such as gerontechnology, biogerontology or geroscience (Kennedy *et al.*, 2014) have emerged, whose only concern is the exploration and validation of such solutions. In addition, significant investments have been made by companies, governments and international organisations in this domain of research and technological development as it is seen to be a route for significant financial returns. This is seen as an investment, enabling the exploration of new opportunities arising from the consolidation of 'silver markets'.

In this regard, research and development on ageing has been underpinned by what Rip (2002) labelled 'strategic science', where attempts are made to align laboratories, policy makers and wider publics with recognised forms of the 'good', such as wealth creation, sustainability or social integration. As a consequence, researchers and technological developers in the domain of ageing are increasingly required to orientate their projects and programmes towards shared societal expectations. This means that knowledge making institutions and tools cease to be solely an expert's domain, and are made more open to extended peer review and evaluation, seeking alignment between research and social needs and values, and where lay users or citizens become directly involved in the innovation process (Nowotny *et al.*, 2001). Indeed, social research has amply documented how and for what reasons lay members are increasingly asked to participate in the articulation of expectations from research, and in the co-production of knowledge more generally (Callon, 1999; Irwin, 2006; Marres, 2007; Callon and Rabeharisoa, 2008). This is also true in the domain of ageing. For example, in the *Research Agenda on Ageing for the Twenty-First Century*, produced by the United Nations and the International Association of Gerontology and Geriatrics (IAGG), it is recommended that dialogue should be underpinned by the use of a 'common language' between researchers, policy makers and publics, and that this 'should lead to a sense of association of *all key players*' involved in research (UNPoA/IAGG, 2007: 11, my emphasis).

However, as the research on social robust science above also demonstrates, the aim to involve publics in innovation and promote dialogue between lay, expert and policy groups is a difficult business (Irwin and Horst, 2015). In the domain of ageing, there are signs that the association or alignment 'of all key players' is difficult to achieve. For example, in the UK Biological and Biomedical Sciences Research Council/Medical Research Council 2006 public consultation on ageing research a gap was identified between, on the one hand, scientists' interest in tracing the ageing process to find conjoint biological roots to downstream age-associated illnesses and, on the other, established cultural perceptions that ageing, *in and of itself*, is not a problem for which a solution is required (Ipsos Mori, 2006; also Irwin, 2006). This finding is indicative of key, wider uncertainties about ageing as an issue in contemporary society, of which the ageing society controversy is but one example. Questions abound.

Is ageing a 'normal' biological process from which pathological states emerge, or is it a condition that requires only prevention and management? Is the aim of ageing research and technological development to erase what we recognise as the experience of becoming older? Or should technology be used only to ameliorate the functional decline associated with ageing? Can the 'negative side' of ageing be erased without consequences for how we experience the later part of life? Can we keep younger and still enjoy the brighter or wiser dimensions of later life experience? What legitimate public expectations should the public have from ageing research? Should we expect to continue to extend our life span, or just to ensure that we are living healthier lives? Can we do the latter without the former? Is healthy ageing an appropriate, achievable aim for all individuals in all societies? There is also uncertainty as to which publics to involve in addressing these societal issues. Is ageing research and innovation only relevant to 'older people' or should a wider public be called into the fold? Is the experience of living longer a key form of knowledge to guide the direction of research on ageing? Does the involvement of 'older people' reinforce age categories and ageism? How are we to identify who should be involved if the accuracy of age grouping is itself questioned by ageing researchers? What criteria other than chronological age can be used to delineate the concerned publics of ageing research?

Ageing thus presents a challenge to contemporary research policies and institutions. There is uncertainty regarding its normative aims and social expectations, the means through which these aims are supposed to be attained and the identity of the beneficiaries of scientific, technological and social innovation. Why, how and for whom should ageing science and technology be supported? My suggestion is that the relevance and significance of these questions are, in no small part, linked to a lack of understanding of the role of technology, science and medicine in the ageing society. Current debates on the ageing society, as I suggested above, tend to view science and technology solely as a solution to the 'problems of ageing'. In this perspective, science and technology are cast in terms of the consequences, the positive – or negative – 'impact' they might have on socio-economic relations, welfare programmes or the experience of old age. However, as most social demographers would agree, ageing societies are themselves the product of changing living contexts associated broadly with industrialisation processes and the establishment of modern science and technology. As a consequence, focusing on science and technology merely as a solution to the 'ageing problem' has undermined our capacity to uncover how science, technology and medicine have been themselves implicated in the making of the ageing society. Investigating this relationship requires more than simple, general models on how science and society are related or should interact. It demands conceptually robust, empirically based, wide-ranging theory. In short, it entails taking the relationship between science technology and ageing seriously.

Concerning science and technology in the 'ageing society'

This book proposes that to take up the challenges set by the 'ageing society' it is necessary to explore the relationship between science, technology and ageing, and that this is best done by drawing on the scholarly traditions developed within Science and Technology Studies (STS). Briefly defined, STS is an interdisciplinary field that investigates the origins, dynamics and implications of science and technology. One of the defining features of this field is that it focuses on knowledge and technology-in-the-making. Rather than looking for rational justifications for ready-made knowledge claims or exploring solely how technology impacts on social relations, STS investigates instead the complex processes through which knowledge and technology are brought together and brought to bear within socially and culturally relevant settings and relations. In this, it emphasises not only how cultural imaginaries and social relations shape forms of knowledge and technological approaches but also how these, in turn, affect and reconfigure socio-economic activities, political institutions and cultural practices.

Although, as we will explore in Chapter 2, within the social studies of ageing, broadly defined, science and technology have remained a lateral, ancillary concern, there is a recent growing interest in the exploration of the relationships between science, technology and ageing. One possible explanation for this late conjoining of the two fields is that ageing not only troubles contemporary scientific and technological practices and institutions, but is also a challenge for STS' core concepts and methodologies. STS is usually concerned with 'new and emerging' technologies, with understanding their social origins and potential implications. Thus, most research on science, technology and ageing has been concerned with outlining the technoscientific configurations of contemporary ageing identities and practices (Joyce *et al.*, 2015). This book takes a different approach. Taking as a point of departure the proposal that science and technology have shaped the institutional apparatus of the ageing society, this book aims to identify the myriad ways in which this came to be so.

The book is, from this perspective, a 'history of the present'. Proposed by Foucault as an alternative to the social science methodological aim to use the past to explain the present, a genealogical 'history of the present' focuses on the multiple, complex links between structures and practices and singular contingent outcomes. It favours descent and emergence rather than origins and causality. As Foucault suggests, the objective of genealogical research 'is not the erecting of foundations: on the contrary, it disturbs what was previously thought immobile; it fragments what was thought unified; its shows the heterogeneity of what was imagined consistent with itself' (Foucault, 1991: 82). This is of crucial importance in relation to the ageing society, on the existence of which, as we saw above, there is wide consensus. To be able to challenge the official versions of the ageing society, Foucault's (1980) attention to specific 'dispositifs' – apparatuses that articulate the normative, political, epistemic, material and technological dimensions of a contemporary issue – is important.

In its attachment to heterogeneity and singularities, the book takes this focus on 'dispositifs' to avoid producing unifying narratives about the relationship between science, technology and ageing. In so doing, it draws heavily from the methodological propositions articulated by Actor-Network Theory (ANT), a specific approach within the social sciences that specifically focuses on the tracing of how associations between people and things come to be shaped or assembled (Latour, 2005). As I will explain in Chapter 2, ANT is an adequate conceptual tool to pursue a genealogical 'history of the present' because it supports the exploration of the different epistemic and/or technological configurations of age and ageing – as well as their emergence, their evolving controversies and uncertainties. ANT emphasises the complex processes by which specific socio-technical 'agencements' or assemblages were brought to bear on the ageing society. As Çalışkan and Callon explain,

> The term agencement is a French word that has no exact English counter-part. In French its meaning is very close to 'arrangement' (or 'assemblage'). It conveys the idea of a combination of heterogeneous elements that have been adjusted to one another. But arrangements (as well as assemblages) could imply a sort of divide between human agents, those who do the arranging or assembling, and things that have been arranged. [A]gencements are arrangements endowed with the capacity to act in different ways, depending on their configuration. [...] Agencements denote socio-technical arrangements when they are considered from the point of view of their capacity to act.
>
> (Çalışkan and Callon, 2010: 9)

Drawing on this focus on the more infrastructural aspects of the ageing society, the book thus challenges the stable, unified versions of the ageing society by tracing its myriad, often contradictory agencements and assemblages. In this regard, the notion of agencement, which is related to that of dispositif or apparatus (Agamben, 2009), supports a more heterogeneous, empirically driven exploration of the 'ageing society'. As the book will further suggest, the focus on the heterogeneity of the 'ageing society' is enhanced by foregrounding the multiplicity and diversity in the relationship between science, technology and ageing, the multiplicity of its agencements. In this regard, it draws inspiration from the work of Mol (2002) and Law (2002) in their aim to map the ways in which, in practice, different realities are brought together or left apart (also Moreira, 2006).

Taking these orientations together, the chapters in the book put forward a version of ageing and of the ageing society as non-coherent realities. Like 'history of the present', non-coherent reality appears, at first sight, to be a contradiction in terms. But like Foucault's genealogical approach, the work of Mol and Law emphasises the patchy character of the realities we inhabit; indeed, they shift our attention to the fissures that hold the 'ageing society' together. Like Foucault's method, theirs might also be seen as '*a revaluing of values*' (Foucault, 1991), that is to say a reimagining of the agencements of our common world. My argument is that if, as was suggested above, ageing is a generative issue from which present worlds and realities emerge, we are better served by concepts that enable the exploration of the uncertainties and unopened possibilities within those worlds. As I will explain in Chapter 2, the book aims to assemble a 'patchwork' story of the relationship between ageing, science and technology. My view is that this approach to researching the issue is not only adequate to describe its shifting dimensions and relations, but also enables the exploration of possibilities beyond the present moment. Because in ageing not everything is stable, decided, settled, a patchwork narrative brings uncertainty and heterogeneity to bear on our reimagining and revaluing of the ageing society.

Because, as I suggested above, existing research on science, technology and ageing focuses primarily on contemporary technoscientific reconfigurations of ageing identities and practices, to build this 'patchwork genealogy of the ageing society' it was necessary to review and link together previously unconnected areas or domains of research, as well as conduct primary research. The bulk of the material used in this book is composed of published documents, that is to say, books, research papers, policy documents, discussion essays, charts, maps, letters, etc., that are available in libraries, electronic databases and sometimes physical or digital archives. The selection of the documents included in the analysis was guided by the strategy of 'following the actors', those being the agencements of the 'ageing society' that are articulated in particular concepts, instruments, technologies or devices which can be tracked across documents (Latour, 1987). This focus on documents and the methodological strategies associated with it is also intimately related to the genealogical approach as outlined above, in that by not aiming to 'unearth' hidden mechanisms of causation to explain the present situation, I am concerned instead with outlining the formation of new assemblages 'that make things work in a different manner' (Rabinow, 2000: 44).

Structure of the book

Taking the conceptual and methodological commitments explored above, the book investigates a diverse array of 'agencements' that, I argue, are key to understanding the relationship between science, technology and the 'ageing society'. To do so, it first contextualises such investigation within the broad fields of social studies of ageing and STS. Bringing them together, I suggest, requires an in-depth investigation of the ways in which ageing, on the one hand, and innovation, on the other, have been drawn upon as background concepts or external factors in explaining either technoscientific change or the changing social organisation of ageing. Foregrounding the constitutive, generative relationship between science, technology and ageing entails replacing unidirectional technological determinist narratives of the ageing society with an approach that emphasises the plurality of ageing technoscience-mediated worlds: a patchwork genealogy that gathers the multiplicities and uncertainties that hold the 'ageing society' together.

In Chapter 3, I analyse the epistemic, technological and political origins of the concept of the 'ageing society'. In it, I argue that the very idea of the 'ageing society' speaks to a perspicuous problematisation of the knowledge infrastructure of industrialised societies. I outline a genealogy of the relationship between the 'population problem' and the economy, articulated explicitly in actuarial sciences in the nineteenth century, and ultimately in the emergence of state-backed old age pensions in liberal democracies. I then suggest that the ageing society is in fact a questioning of the knowledge base of the institutions of such a liberal welfare state, a process whereby new technoscientific promises – relating to the technological enhancement of health and activity – are articulated to address the failure of demographic calculations and economic forecasting to accurately predict the extent and consequences of increased life expectancies.

In Chapter 4, I focus on the evolution of the problem of declining or ageing populations within population science itself. I explore how the stabilisation of the link between the mechanisms of life and the economy, broached in the previous chapter, relied upon what I label a eugenicist agencement focused on the 'quality of population'. I argue that this concern with 'race' and migratory flows had a significant role in shaping social security systems in liberal welfare states. Focusing on how population science experienced a shift away from lowering birth rates – which it considered key to producing economic development – and towards a focus on 'vitality' and life expectancy, I suggest that the consolidation of the 'ageing society' is underpinned by an interrelated increasing erasure of the role of migration in social renewal. By aiming to make a 'better use' of a settled, declining population, populations science participated, I argue, in a form of political and epistemic exclusion that has deep consequences for current debates on security, race and identity in the contemporary 'ageing society'.

Chapter 5 explores how the widespread belief that there is an inexorable inverse relationship between ageing and productivity or 'vitality' was put into question by a collective investigation on the procedures and methods of age measurement. I describe how epistemic and normative uncertainty about the role of chronological age – the number of years lived since birth – in ascertaining age-specific norms, values and expectations emerged as a critique of the formal rationality of the modern bureaucratic state. This led to a proliferation of intended 'personalised' age standards, developed and highly debated across a variety of forms of knowledge, decentralised organisations and purposes. I further argue that uncertainties in the validation and implementation of these alternative standards have reinforced two modes of organising knowledge production to meet the challenges of the 'ageing society': one, that relies on chronological age (CA) as a variable to construct niche clusters of health risk, and on biomedical expertise as a guide to interpret data and guide individual conduct (Rabinow and Rose, 2006), and another which aims to embody in the alternative standard itself the mechanisms of individualisation of health, work and technology.

In Chapter 6, I extend this analysis of the role of individualising standards in the 'ageing society'. I ask how it is that we have come to understand ageing as an individualised process, resulting from the combination of genetic programmes, psychological characteristics and the evolving interaction with the social and material environment. I argue that this consensus was assembled in and through a particular methodological approach to the study of ageing and development, the longitudinal or cohort study. I trace the co-productive relation-ship between ageing as an epistemic object and the longitudinal study from the early twentieth century by focusing on what is considered the most iconic and successful of such studies, the Baltimore Longitudinal Study of Aging (BLSA), and describe how it developed through the interaction of two diverging know-ledge repertoires: one focused on ascertaining a 'true value' of ageing through the construction of averages, and another reliant on the construction of model physiological relationships. This tension, I suggest, helps explain further the paradoxes of individualisation and of the life course explored in the previous chapter.

Chapter 7 is concerned with how economic activity in later years is seen to be the key for many of the economic problems of the 'ageing society'. Revisiting some of the themes and issues explored in Chapters 2 and 5, I investigate how policy solutions being proposed for the ageing workforce and rising old age dependency ratios are underpinned by instruments aiming to measure the fit between individuals and working technology. It describes how industrial geron-tologists attempted to decouple ageing from declining work ability by articu-lating the instrumented concept of 'functional age'. This concept however was, I argue, not a unitary tool but included two versions of ageing at work: one that used exercise physiology to locate the differential ability to maintain 'internal equilibrium' at the heart of functional trajectories, and another using simulations of actual working environments to understand how to stipulate the fit between average functional declines and work tasks.

Chapter 8 is focused on care as a sociotechnical practice. It departs from the proposal, voiced by an increasing number of policy makers, that older people should be able to 'age in place', to continue living in their homes and com-munities, and that this is best achieved through the design and incorporation of communication and assistive technologies. I suggest that the alignment between in place and active ageing is contingently enabled by a match between how social psychological models of ageing enact behaviour as malleable, and the arti-ficiality of conduct in the neoliberal governing of the ageing society. I argue however that this technoscientific vision is also sustained by making invisible the routine ways which integrate the more mundane technologies in current use into everyday life, and that this questions the sustainability of the regime of tech-noscientific promises upon which the ageing society is founded.

In Chapter 9, the longest of the book, I address the key issue of the relation-ship between contemporary medicine and ageing. There, the opening is provided by current concerns with an increasing medicalisation of old age, i.e. with trans-forming normal signs of ageing into medical problems, and with the anxiety that this is more pressing with the application of biomolecular knowledge and tech-nologies to ageing. While this is an attractive narrative, both for critical practi-tioners and social scientists, I argue that it fails to acknowledge the persistent troubles experienced in trying to establish channels of translation between the biology of ageing – sometimes called biogerontology – and clinical medicine since the middle of the twentieth century. I suggest that, recognising this problem, biogerontologists have built a critical attack on the epistemic and institutional foundations of biomedicine and its inability to deal with age-related illness. This, however, raises uncertainty about the normative goals of technological intervention in the ageing process, and whether it is possible or desirable to maintain health for longer without extending the human life span.

In Chapter 10, the conclusion, I identify key running themes across the chapters to question and explore the idea that ageing and the ageing society might be becoming a 'thing of the past'. The chapter proposes that an epistemic understanding of the present ageing society is crucial to imagine different articulations between science, technology and the 'unfolding of life' but in order to so it is necessary to 'gather in tension' multiple enactments of the ageing assemblage.

Chapter 2

Patching the science, technology and ageing conjunction

Introduction

In the last chapter, I proposed that to understand the ageing society we should focus on the knowledge making practices, tools and technologies, and the knowledge making institutions that underpin it. The suggestion was that the challenges set by the ageing society are best addressed by focusing not only on the demographic, economic and socio-cultural consequences of ageing, but specifically on the relationship between science, technology and ageing. I argued that whereas science and technology are usually seen as solutions to the issues brought to bear by the ageing of populations, science and technology have been implicated in the making and managing of the ageing society from the outset.

Once we focus on the knowledge infrastructure of the ageing society, all that appeared solid and stable turns uncertain and contested. This means that instead of finding in science and technology the unwavering foundations upon which a new organisation of the ageing society should be built, we find more questions than answers. These questions, I argued, are not an accident: they are constitutive of the ageing society itself. Taking these questions seriously entails researching the role of science and technology in society. This is best done by drawing on the scholarly traditions developed within Science and Technology Studies (STS).

In this chapter, I explore how STS can be deployed to understand the ageing society. The chapter critically reviews an emerging literature on science, technology and ageing to argue for a perspective that is methodologically committed to opening up the social and technical worlds of ageing. Rather than critically juxtapose present situations with idealised forms of getting older, I suggest that it is possible to use STS concepts and methods to show how 'it could have been otherwise' and also bring them to bear on the ways in which it has *been* otherwise. Such emphasis on the composite, the multiple – on 'the heterogeneity of what was imagined consistent with itself' (Foucault, 1991: 82) – requires some attention to the conceptual apparatus used in this book.

This conceptual investigation is necessary because it makes public circulating but often tacit assumptions about the nature of the ageing process and how it relates to scientific practice and technological innovation. In particular, it enables us to interrogate the centrality of agency in the relationship between science, technology and ageing. In a context where older people are increasingly urged to engage with science and technology in order to enhance their health and involvement in economic and cultural spheres, where 'active ageing' is the order of the day, it makes renewed sense to investigate how agency and activity are related in theories about innovation and ageing. However, the chapter proposes that rather than investing in a generic attribution or recovery of agency in older people's engagement with technoscience, we should be more sensitive to the less than coherent way, the sometimes inconsistent way, in which power is distributed in these engagement processes. This, in turn, means we must pay special attention to the tropes, the literary devices we use to describe the nexus between science, technology and ageing, looking beyond the literature that explicitly focuses on such a topic.

The chapter first addresses the more contemporary explorations of this nexus to then investigate how two main figures have dominated academic thinking about technoscience and ageing: one emphasising unidirectional, technologically determined changes in the nature and experience of human longevity and ageing, and the other viewing technology as having differentiating, phased effects on those experiences. The chapter recommends the figure of the patchwork as a way to think through the pragmatic, emerging, diverging complexity of the nexus between technoscience and ageing.

Science, technology and ageing

One of the key proposals in this book is that STS can make a useful contribution to our understanding of the ageing society. But what is STS and why is it useful? Briefly defined, STS is an interdisciplinary field that investigates the social, cultural and political origins, dynamics and implications of science and technology. Tracking its origins to emerging public concerns about the role of science and technology in economic, social and political processes in the 1960s, STS acquired a distinctive epistemic identity in the 1970s when academic researchers began articulating the specific conceptual and methodological issues raised by wanting to understand science and technology in their own terms, and not merely as secondary phenomena. In this process, STS has come to encompass diverse approaches such as the economics of innovation, sociology of scientific knowledge, research policy, public understanding of science or cultural studies of science and technology.

Despite its internal diversity, one of the defining features of this field is that it focuses on knowledge and technology-in-the-making. Rather than looking for rational justifications for ready-made knowledge claims – as philosophers of science would do – or exploring solely how technology impacts on social relations – as mainstream social scientists tend to do (see below) – STS investigates instead the complex processes through which knowledge and technology are *brought together with* social and cultural relations. In this, it explores not only how cultural imaginaries and social relations shape particular forms of knowledge or technological approaches, but also how these in turn affect socio-economic activities, political institutions and cultural practices. To conceptualise this interweaving of science, technologies and social, normatively embedded practices, STS scholars have proposed the notion of sociotechnical relations, agencements or assemblages (Chapter 1).

The appeal of these concepts is to highlight the ways in which what we usually conceive of as social relations – families, firms, schools, clubs, etc. – are embedded in technological infrastructures and reliant on specialised forms of knowing (Latour, 2005). Conversely, it draws attention to the often unexplored moral, cultural assumptions or social, pragmatic implications of scientific innovations or technologies (Akrich, 1992). A sociotechnical assemblage or agencement is an open-ended gathering of several different, sometimes in contrast, elements that can range from material artefacts to texts, to people or organisations. The relations between the elements of the assemblage are non-linear, complex and never fully predictable. In an assemblage, components relate to each other as contingently obligatory elements (Callon, 1986). They can, however, be traced and their dynamic transformation described and understood.

While the notion of the sociotechnical became established as a central epistemic commitment in the field of STS in the mid 1990s, it was only at the turn of the century that its implications for mainstream social sciences began to be spelled out. Scholars in disciplines such as anthropology, sociology or geography, particularly in Europe, began to question the suitability of a generalised – usually negative – view of the role of science and technology in the making of the social, and explored the possibility of opening their analytical lens to sociotechnical processes. Equally, in applied fields such as health research, information science or marketing, the ability to take into consideration the contribution of knowledge or artefacts in health care or computing became a welcomed possibility.

In social gerontology and sociology of the life course, applied fields concerned with the social and cultural dimensions of the ageing process, however, science and technology have remained a lateral, ancillary concern in terms of volume of research and research focus. As a result, there has not been a sustained engagement with the notion of sociotechnical relations. This is not to say that social gerontology has not engaged with the 'question of technology' (e.g. McCreadie, 2010). Indeed, as we will see in the next section, the nexus of technoscience and ageing has played a key role in shaping social scientific thinking about innovation, on the one hand, and the life course, on the other. Further, from the turn of this century onwards, diffusion of information and communication technologies into the domestic environment, often as a possible means to provide health and social care to older people (e.g. Loader *et al.*, 2009), has slowly emplaced the 'question of technology' within social gerontology's problematic. This is particularly visible in the consolidation of the field of gerontechnology – research to develop and understand technological applications for older users (Bouma *et al.*, 2007) – in those same decades.

This is not to say that STS has been in any way more open to the issue of ageing. Taking publications in the journal *Social Studies of Science* as an indicator, ageing is solely explored while the field is still emerging, in the context of understanding 'age stratification' in productivity or peer recognition (e.g. Stern, 1978). It will take more than two decades for a renewed link between ageing and STS to be forged, this time focused explicitly on the technoscience of ageing (e.g. Fishman, 2004). This later body of research draws heavily on the concept of sociotechnical relations or assemblages, aiming to understand how science and technology approached, shaped and transformed the meaning and experience of old age, tracing the technologically and scientifically mediated changes to the material experience of later life (Peine *et al.*, 2015).

In shaping this movement of bringing ageing into the STS research agenda, Joyce and Mamo's 'greying' of the cyborg is of key importance (Joyce and Mamo, 2006). Drawing on Haraway's (1991) conceptual figure of a human–machine chimera designed to support thinking beyond the prevailing dichotomies of male–female or nature–culture, Joyce and Mamo suggest that the power of Haraway's notion can be extended through questioning the young–old opposition that usually shapes discourse around technology design and (non) use. This aim, in turn, relies on bringing the notion of sociotechnical relations to the fore to understand 'the complex ways technologies and sciences constitute the meanings and experiences of ageing as well as the ways ageing people negotiate and give meaning to these in their own lives' (Joyce and Mamo, 2006: 100; see also Joyce *et al.*, 2015). Like Haraway before them, Joyce and Mamo are committed to the possibility of redeploying the conjunction of ageing, gender and technology.

In what constitutes the first explicit articulation of the value of STS for the study of ageing, Joyce and Mamo aim to do this by focusing on the 'agentic' capacities of ageing people to negotiate how medical categories and health and care technologies are deployed in their everyday lives. This they see as a departure from the more structuralist approaches to ageing and 'old age' – often associated with critical gerontology – whereby knowledge formations are seen merely to reinforce existing inequalities, and thereby limit the capacity of older people to resist social and political exclusion. Similarly, Peine *et al.* insist that recognising the agentic capacities of older people 'throws into question existing stereotypes of older persons as passive recipients of science and technology' (Peine *et al.*, 2015: 3). In this respect, both their proposals can be seen as furthering Czaja and Barr's call 'to reconceptualise older individuals as active users of technology rather than as passive recipients' (Czaja and Barr, 1989: 128). But whereas the latter aimed simply at designing better, more inclusive technologies, STS scholars suggest that we should sharpen our 'attention to the way old people approach technology, science, and health in creative ways' (Joyce and Loe, 2010: 173). Going beyond the aims and objectives of designers, older users not only reinvent the purpose of the tools they use but also, by the same stroke, pragmatically contest stereotypes of old age and recast the meaning and experience of later life.

There appears to be a commitment among scholars concerned with applying STS to ageing to repairing, in their approach and analysis, ageing individuals' ability to affect the material environments and institutional contexts in which they live. Extending a strict sociological definition of agency to a wider, more heterogeneous conception of the elements that partake in and shape people's lives, this restitution of power is motivated by the sense that older people are normally construed as 'passive recipients' of science, technology or biomedical innovation. In contrast with technoscience's and biomedicine's assumptions about older people's impassiveness and disengagement, STS scholars urge us to provide rich and detailed accounts of the imaginative ways in which older people interact with knowledge making institutions and technological processes (e.g. Mort et al., 2015; Östlund et al., 2015). STS-inspired analyses of older people's lives can redistribute agency and powerfully deliver innovative normative, ethical grounds for practice, rearticulating imaginaries that refocus the aims of science and technology in an 'ageing society'.

Where this request for agentic redistribution is less appealing is in how it overlaps with contemporary calls for a fuller and more encompassing, active involvement of older people in society. Indeed, as I indicated in the previous chapter, policy responses to demographic ageing have, in the last two decades, emphasised the need to create an age-integrated society, moving towards initiatives that promote what is often labelled as 'active and healthy ageing'. These are conceived as strategies that promote the extension of healthy life expectancy and remove institutional barriers to older people's economic and social participation (Fernández-Ballesteros et al., 2013). Policy analysts and critical social gerontologists suggest however that such programmes shift normative expectation regarding older people and, in particular, emphasise the extension of active working life in detriment of other aspects of wellbeing. Thus, active ageing policies have become the target of criticisms for potentially discriminating against the 'passive' and dependent, and for their inability to integrate the experience of decline or loss of function (Moulaert and Paris, 2013).

Some scholars have linked the emphasis on activity and health to the consolidation of neoliberal, biopolitical forms of governance (Katz, 2000; Katz and Marshall, 2003), whereby power is deployed through the forms of knowledge that individuals produce in the maintenance of their own bodies (Rose, 2009). In a study of media representation of active ageing in Canada, for example, it was shown that the individual is consistently identified as the key agent for a successful ageing process and a functional ageing society (Rozanova, 2010). In our own study of the transformation of biological models of ageing, Palladino and I have argued that the contemporary, dominant focus on 'the evolved capacity of somatic cells to carry out effective maintenance and repair' (Kirkwood and Austad, 2000: 235) is redolent of the,

transformation of socio-political forms of management whereby [...] the individual of the 19th-century biopolitical imaginary – a human body whose biological constitution was irremediably fixed at birth – is giving way to an understanding of the human body as an assembly of biomolecular components that can be [...] recombined so as to maximize the resultant unit's cultural, social and political productivity.

(Moreira and Palladino, 2008: 21)

While it is in the discourse of healthy ageing that this reconfiguration of activity in later life is most visible, its sociotechnical and biopolitical penetration – the way it shapes older people's lives – should prompt our rethinking of the role of agency in the ageing and technoscience nexus. Can our call for better accounts of older people's engagement in sociotechnical processes be reinforcing the moral and techno-political imperative to be active, involved, healthy and productive? A similar paradoxical effect of attempting to redistribute agency within contemporary enactments of biopower was examined by Callon and Rabeharisoa (2004) in their well-known 'Gino' paper. Tasked with investigating forms of patient and public involvement with biomedicine in the rare disease domain, Callon and Rabeharisoa were puzzled not only by their interviewee's refusal of any engagement with genetics or biotechnology, but especially by his refusal to account for this 'choice' or decision. Yet, in attempting to understand Gino's silence, and in using social science's prime technique – the interview – to make his desire for opacity visible, Callon and Rabeharisoa could only enrol him in the very network of accountability that he was carefully avoiding. In bringing to bear his reasons for non-engagement, and transforming it into a form of agency, Callon and Rabeharisoa are, by their own admission, paradoxically working against the non-accountability Gino would appear to desire most.

It is not a coincidence that this same effect impacts on theorising about agency in the ageing and technoscience nexus, as we draw from the same corpus of social research categories, techniques and methodological devices. As Callon and Rabeharisoa (2004) suggest, the social sciences are epistemically and normatively committed to multiplying and accounting for 'local' ways of being, embedded in tradition, structural conditions or knowledge of the social world. While agentic accounts of social action tend to be associated with the latter, all forms of social science theorising, lay or professional, constitute a distribution of agency because they credit or discredit particular actors or components 'in the accounts they provide about what makes them act' (Latour, 2005: 52). In our attempts to understand the 'world of older people', and account for their active role in shaping it, we also enact a particular distribution of agency, one that, perhaps paradoxically, morally deflates the possibility of being passive in later life.

One possible way out of this paradox would be to foster a different form of theorising the ageing and technoscience nexus, one that emphasises uncertainty, multiplicity and complexity. This is something that STS has been very good at, by exploring the contingencies, the manifold uncertainties that underpin sociotechnical processes, and in providing insight into how the – often partial – resolution of these uncertainties affects subsequent negotiations and conflicts. This, in turn, has been key to opening the imagination to multiple, possible paths not taken. STS' peculiar imagination is underpinned thus by the acknowledgement that 'it could have been otherwise'. A fundamental consequence of this openness is that rather than taking as its point of departure a stabilised definition of the issue under consideration – say ageing, in our case – and of the range of conditions or variables that are relevant for the understanding of that issue, STS scholars are exactly concerned with exploring how the issue came to be articulated in a particular way, and trace what and how a myriad of elements became entangled and transformed in this process of co-production. This mutual entanglement between the issue and what it relates to means that causality pathways can be reversed and confused, and that what started out as a driving, leading, agentic force might, in the process, become a weaker, dominated element in the assemblage (Callon and Latour, 1981; Callon, 1986; Latour, 1987, 2005; Callon and Law, 1995).

This might be seen as a radical proposition, in that it aims to substitute theory, in its usual sense, for detailed, conceptually driven descriptions of how the world is built and explained by those entangled, in some way or another, in the building of it (Callon, 1986; Latour, 2005). Primarily, adopting this approach is a methodological device to support the exploration of sociotechnical agencements, with different epistemic and/or technological arrangements of age and ageing. It enables understanding their genesis, their evolving controversies and uncertainties, as well as the partial stabilisations of elements and the entrenchment of standards and modes of explanation. It also supports tracing how sometimes such worlds are deposed, and explaining their eventual demise. Crucially, these devices care for the mapping of heterogeneities that reverberate in knowledge making institutions, technology and biomedical projects as they entangle themselves with the issue of ageing. It facilitates, to reiterate Foucault's words, our writing about 'the heterogeneity of what was imagined consistent with itself'.

This perspective – sometimes labelled cosmopolitical (Stengers, 2005) or 'multiplicity realism' (Zuiderent-Jerak, 2015) – not only points to how 'it could have been otherwise' but more importantly to how *it is otherwise*, always, already. First articulated in the work of Annemarie Mol (2002; Berg and Mol, 1998) and John Law (2002; Law, 1994; Law and Mol, 1995, 2002), this approach has been concerned with extending the materially heterogeneous focus of STS, explained above, to address the challenges of studying coexisting, parallel sociotechnical assemblages: different ways of diagnosing and treating an illness (Moreira, 2006; Moser, 2008), or disparate ways of enacting the value of food (Mol, 2012), for example. This has consequences for how we theorise the nexus between ageing, science and technology, and the different distributions of agency we find in the controversies about that relationship in the contemporary world. While a customary social science approach would seek to rank these differences through first principles or theory, the proposal here is to map the ways in which, in practice, those modes of existence are brought together or left apart. This means that it brings to bear a politics of method that explicitly challenges some of the assumptions about the role of social science in the world. It asks 'what would happen if we abandoned the idea that [stories] describe a single set of [realities] and instead [take up] the idea that they are performing different distributions?' (Law, 2002: 158).

This is exactly what Palladino and I (2008) have aimed to do in our analysis of biological models of ageing, referred to above. While acknowledging the significance of neoliberal, governmentalised modes of getting older in scientific practice and technological innovation, we traced how these coexisted with other versions of the nexus between ageing, science and technology. In the same vein, Lassen and I (2014) have traced the multiple ways in which 'active ageing' itself is enacted and how they relate to divergent distributions of agency for older people. Drawing on this body of work, the chapters in this book explore different distributions in the science, technology and ageing relationship. What emerges from this is an understanding of ageing as a less than consistent sociotechnical assemblage, because the different distributions do not fully hold together. By this I do not mean that ageing is a wholly disjointed topic of collective investigation, but rather that the links between the issues and topics analysed in different chapters are contingently obligatory and emergent, that is to say, they are composed of a multitude of different assemblages that are only connected in partial, fragile ways.

To reiterate, whereas conventional methods of social science tend to erase this incomplete coherence, premised as they are on the assumption that reality is singular (Law, 2004, 2015), the ability to trace heterogeneities relies on deploying ageing as a generative issue, leading to a proliferation of relevant worlds and to the opening of uncertainties. Such a conception of ageing enables the exploration of possibilities, of *the otherwise*. Instead of an effort to 'bring together' and summarise analyses, we acknowledge the myriads of 'forces', alignments, etc., at play in any particular agencement, to explore the heterogeneities that have brought ageing to bear in contemporary societies. We thus should not aim to identify a single technoscientific story or narrative, or even two, through which the meaning and experience of ageing can be explained or understood. We should pursue many partially connected stories of the ageing assemblage. But how can this be done?

My suggestion is that this requires exploring the often hidden ways in which we *account for* the nexus between science, technology and ageing. By critically examining the tropes, the rhetorical and story-telling devices that are used to describe this relationship, it then becomes possible to tinker and modify them to suit the problem at hand. For this, however, we have to go beyond the recent literature on science, technology and ageing, and unearth this relationship in research on the social organisation of ageing, on the one hand, and innovation processes on the other. While not being acknowledged as a central component of the explanatory models of these theories, the ageing and technoscience nexus, I will argue, is nonetheless essential to the logic of the explanation. This will be the basis upon which I will propose a different literary device – the patchwork – to account for this relationship.

The nexus of ageing and technoscience

The previous section established that the relationship between ageing and technoscience has been under-investigated both in the field of social gerontology and
STS. Where it has become the focus of research, this has been underpinned by a
concern with recovering agentic qualities for older people, which are seen to be
absent in the normative scripts embedded in age-related technoscientific facts
and artefacts. I have argued that such a focus rubs uncomfortably with dominant
discourses on active and healthy ageing, which aim to undo the passive way
older people have previously been cast in and through policy instruments. I have
suggested that instead we should open our investigations to multiple, coexisting
ageing assemblages, each one 'performing different distributions' of agency. To
do this, however, we need to look carefully into the way in which the nexus of
technoscience and ageing has been described, often in an ancillary way, in literature about ageing in society, on the one hand, and industrial innovation, on the
other. Looking at these bodies of literature, my contention is that there are two
main tropes through which the nexus between ageing and technology has been
deployed: arrows and waves.

Arrow models see the relationship between ageing and technology as caught
in a unidirectional movement whereby scientific and technological rationality
comes to shape the age-relevant institutions and social practices. Some versions
of this approach, as we will see below, embrace technological determinism – the
idea that scientific and technological innovation controls the direction of social
change – while others emphasise how social and political factors themselves
shape research and innovation. Some see rationalisation, modernisation and
standardisation as a force for good, emplacing social and cultural conventions on
grounds of objective truth, while others see these processes as undermining sedimented traditions and local knowledge.

Wave models conceptualise the relationship between ageing and technology
as an oscillatory movement. Scientific and technological innovation is characterised by different phases which are associated with particular forms of social and
economic organisation. Life course processes are shaped by these interactions
between technology and society. Wave models also come in different varieties,
some emphasising the economic driving forces of change, while others
emphasise the life course consequences of innovation processes.

One illustrative example of arrow models can be found in the work of Lawrence Frank, a social scientist who, between the 1930s and 1950s, played a key
role in shaping gerontology as a field of research and social gerontology in particular (Bryson, 1998; see Chapters 5 and 6; see also Achenbaum, 1995). Frank
never became a full academic sociologist, working primarily in applied social
research before being appointed in 1923 to direct research into child development at the Laura Spelman Rockefeller Memorial Fund. His approach to social
science was shaped by the cultural and normative consensus of the Progressive
Era (1890–1920), where the social and economic sciences became integrated in

the apparatus of policy making of the state. For Frank, the social sciences represented a necessary tool to respond to the challenges of modernisation. In particular, he was concerned that a transition from homogenous to differentiated societies had led to moral fragmentation. This process was, according to Frank, at the core of America's social problems – crime, delinquency, unemployment, etc.,

> arising from the frantic efforts of individuals lacking any [...] guiding conception of life, to find some way of [...] merely existing on any terms they can manage in a society being remade by technology.
> (Frank, 1936: 339–340, my emphasis; see also Frank, 1925)

Such a conception of the effects of technology on the moral order is characteristic of early sociological thinking, being present also in the works of Emile Durkheim in a European context. It proposes that the application of reason to production – the industrial division of labour – has led to the reorganisation of social relationships, whereby roles have become differentiated. As in Durkheim's conception of anomie, Frank believed that such a transformation had created a divergence between the economic organisation of society and its cultural and normative forces, leaving individuals with no 'guiding conception of life', and thus making individuals act only in their immediate self-interest. His view was that the social sciences, based on these premises, should underpin policy programmes and reforms in attempting to align moral imperatives with the structural demands, creating what he labelled new 'designs for living' (Frank, 1939: 344).

Frank's transference to the Macy Foundation coincided with the establishment of the Social Security Act of 1935, which created assistance and insurance programmes for 'old age' in the United States. Having been pivotal in setting up a network of research institutions that shaped scientific understanding and measurement of child development, Frank understood his new position to be a similar one, and centred his work on outlining how divergences between biological, psychological, economic and cultural processes were at the basis of 'the problems of ageing'. In making this case, Frank argued that there was a need for a new science of ageing – gerontology – because,

> [s]tatistically there are millions of men and women growing older, all more or less baffled by the changes within themselves and the confusion and turmoil in the world around them, presenting much the same fundamental needs and problems in the aggregate for which large scale provisions must be made. Yet, each one is a distinct, unique organism-personality, with his or her idiosyncrasies of functioning and living, with his or her idiomatic way of acting and feeling [...]. This, then, is the test of our many organizations, institutions, professions, and services which the pressure of technology and of standardization is driving toward a large impersonal rigidity that

may become more terrifying than the ills, hazards, and insecurities against which they are organized to defend us.

(Frank, 1946: 9)

For Frank, modernisation presented two combined challenges. On the one hand, the acceleration of change in customs and ways of living compounds the biological and psychological changes that older people experience in their lives, leaving individuals 'baffled' and confused. On the other, 'technology and ... standardization' drive welfare programmes to address those individuals with 'impersonal rigidity', which makes the experience of growing older more difficult and distressing. The 'problems of ageing', then, are, for Frank, a consequence of the ways in which modern, technological change is at odds with how – what he called – individual organism-personalities become older (see Chapter 5). In its drive for technical efficiency and rationalisation of production, modernity disregards the individual ageing process, contributing to a sense of social and cultural exclusion in older people. What is interesting in Frank's analysis is that while he does not question the arrow of technological progress and modernisation, accepting the economic and social gains it has brought, he is still engaged in finding ways to mitigate its nefarious effects, and in particular how it leads to 'old age' being characterised by anxiety, dis-adjustment and insecurity.

Such a view of the effects of modernisation on the ageing process and experience encountered a fuller academic development in the work of Howard Cowgill. Cowgill's work was underpinned by an interest in exploring the relationship between demographic and social change, extending the models proposed by demographic transition theorists (e.g. Davis, 1945; see also Chapter 4). Demographic transition theory suggested that modernity had been accompanied by a transformation in the structure of populations, from those shaped by a high number of births (fertility) and high mortality to more recent cases where both death and fertility rates decline. One of the consequences of this transition is the ageing of the population, as the number of births declines but life expectancy increases. But while demographic transition enabled an understanding of the societal consequences of demographic ageing, it did not explain why modernisation 'tended to devalue old people and to reduce their status' (Cowgill, 1974: 10). Cowgill argued that, along with education and urbanisation, technological change created the conditions for such devaluation in two related ways.

First, the introduction of sanitation and hygiene procedures helped drive down infant mortality rates (see also McKeown, 1979). In a second phase, implementation of health technologies contributed to prolonging average life spans. These combined to create a permanent 'surplus' of labour and the creation of retirement as 'a social substitute for death' (Cowgill, 1974: 12; see also Achenbaum, 1978), leading to the increasing segregation of older people within a work-oriented culture. In addition, the introduction of,

modern economic technology [...] creates many new occupations and transforms most of the old ones. It is only natural that the people coming forward
to fill these new jobs should be those not yet established in careers – namely,
the young. [...] Older workers carry on in the more traditional work roles,
some of which become obsolete and most of which are less highly valued
and, therefore, less well remunerated.

(Cowgill, 1974: 13)

Cowgill's model hinges on the tension between two forms of technology, one
which prevents death and the other seeking new, more efficient ways to produce
wealth. As in Frank's work, 'old age' is portrayed virtually as a by-product of
larger technological changes, except here the subjective frustration and dis-
adjustment is mostly explained by social processes of value-attribution, that is to
say, by the fact that society prizes economic activity and efficiency. As in
Frank's work, technological innovation and diffusion are here also taken for
granted, their social origins left unexplored and only their consequences outlined. Indeed, accepting uncritically the desirability of modernisation as a
process has been one of the main criticisms levelled at both demographic transition theory and the application of modernisation theory to ageing (Turner,
1989). The main issue with this uncritical approach, from our perspective, is that
scientific progress and technological change come to be seen as inescapable,
relentless processes that shape society's institutions and practices.

To an important degree, analyses such as these were challenged by the emergence of a post-modern critique of grand narratives of social and scientific progress (Lyotard, 1979) and more specifically by the emergence of critical
gerontology at the end of the 1970s (e.g. Estes, 1979). Critical gerontology
directly challenged the processes of commodification and bureaucratisation of
'old age', seeing them not as a side-effect of modernisation but as a direct consequence of policies and professional practice aiming to control older people
(Townsend, 1981). Although not focusing specifically on scientific or technological innovation, critical gerontology suggested that gerontology itself partook in
the 'social engineering' approach to older people (Moody, 2006), denounced the
capitalist economic roots that shaped such an approach and the 'designs for
living' it proposed. These challenges to mainstream gerontology, initially
inspired by Marxist political economy, were then to be further enriched by feminist, post-modern and Foucauldian gerontologies. Feminists emphasised the
way in which power imbalances along gender lines have shaped knowledge formation, the expectations about and policies for older people (Arber and Ginn,
1991). Post-modernists highlighted the increased focus on technologies to
modify – to rejuvenate – not just bodily appearance but the materiality of the
body itself (Featherstone and Hepworth, 1991). Foucauldian analyses encased
the development of knowledge about 'old age' within a process of discourse formation intimately linked to forms of 'control-at-a-distance' to shape individual
behaviour and manage populations (Katz, 1996).

However, despite key differences in both functionalist and critical perspectives, the relationship between technology and ageing is deployed through distinct, overarching historical, teleological processes. The main consequence of positing a cohesive 'march of history' is that exploration of tensions and alternatives can only be done through either a reparative, social engineering approach such as Frank's, or by attempting to rebuild the very foundations of the socio-technical organisation of ageing by, for example, reinstituting the agentic capacities of older people in shaping the worlds in which they live (see previous section).

Another way of thinking about science, technology and ageing draws on the imagery of waves, and is most easily identified in studies of innovation processes. Famously, the economist Joseph Schumpeter defined capitalism as 'an evolutionary process' characterised by internally generated change that creates qualitative, discontinuous changes. The driving force of this change, according to Schumpeter, was innovation: the creation of a new product, production method, form of organisation, or the related establishment of the new market. The innovation process, Schumpeter argued, 'incessantly revolutionises the economic structure from within, incessantly destroying the old one, incessantly creating a new one', adding that 'this process of creative destruction is the essential fact about capitalism' (Schumpeter, 1950: 83).

Challenging the assumptions about the equilibrium of the economic system in neoclassical economics, Schumpeter suggested that economic development was dependent upon repeated waves of 'creative destruction' when innovations accrue competitive advantages in the market. Interesting, from our perspective, is the fact that such innovations are conceived as generated within the economic structure leading to a differentiation between 'new' and 'old' firms, ways of working, etc. This model thus links categories of old and new to contingent historical circumstances. Old and new are understood as relative, changing characteristics of objects and people. We are all familiar with this process from using information and communication technologies: what at one point is considered a cutting-edge, fully functional device (phone, tablet, etc.) becomes very quickly obsolete due to a 'new generation' of processors or a new arrangement of components being introduced into the market. These are also markers of people's age status, familiarity with particular forms of computer software or applications being a sign of belonging to the consumer-led, 'culture of the present'.

These processes are well explored by Nowotny, in her analysis of the emergence of new forms of temporality linked to the consolidation of the information and communication technologies in the 1980s:

> Ageing is measured on a socially accepted scale of reference, which is generally the chronological order of calendar time. But this level of abstraction is not very meaningful when it is a question of grasping the social and individual differences which come to light. Above all, the cultural definition intervenes which distinguishes ageing from 'becoming obsolete'. Becoming

obsolete means ageing more quickly than is to be expected from the average rates of ageing or is prescribed by the norm. If technologies, firms and regions age more quickly and clearly 'become obsolete' today, just like people's professional qualification [...], then it seems reasonable to suppose that the norm of ageing changes according to the amount of innovation. The faster the rate of innovation, the faster the growth of proneness to obsolescence.

(Nowotny, 1994: 65–66)

Nowotny contrasts a 'socially accepted', established system of age grading with one that is relative to the 'rate of innovation' (see Chapter 5). Technologies, firms and people age, in Nowotny's definition, because they 'become obsolete', outdated, superseded by another, newer assemblage of technologies, practices and institutions. This process was already hinted at in Cowgill's analysis of the effect of economic technology on ageing, above. What is different in Nowotny's analysis is that the innovation is conceived as a variable condition in the production of age and, importantly, that age is linked to the relationship between two set of technologies, institutions and practices. This proposition is also different from what gerontologists usually label as generational or cohort effects, to refer to the impact changing social conditions have on the lives of individuals born in the same space–time context (e.g. Britain in the post-war years, or 1980s America). It is true that individuals born in the 1970s in most of the Western world would have seen the advent of colour television and the personal computer, and most would have become familiar with handling and working with those devices. But, while in a 'cold' innovation context those technological practices – and correlated identities – would have remained up-to-date, in a 'hot' innovation context they quickly become obsolete.

Changing sociotechnical conditions, Nowotny seems to be suggesting, affect the way in which we experience life course processes. This is because of how we, and our lives, are attached, entangled with particular technologies, in what Callon (1992) describes as techno-economic networks. Through this concept, Callon makes the link between evolutionary economics and innovation studies, on the one hand, and science and technology studies, particularly Actor-Network Theory (see above), on the other. For him, innovation and market creation are outcomes of distributed, tentative processes experimentation, only 'some of which give rise to market transactions' (Callon, 1992: 73; Callon, 1998). The solidity or robustness of those networks is variable and relative to the emergence and establishment of other technologies in the same domain, as Nowotny suggests. For example, some technologies such as the internal combustion engine powered vehicle have not suffered fundamental alterations since it began being mass produced by companies like Ford, whereas the use of magnetic tape to record and store motion images and sound (aka videotape) is now mostly an archive resource or museum piece. Some sociotechnical assemblages are 'locked-in' in processes of reinforcement and irreversibility (Callon, 1991;

Arthur, 1989), while others are volatile and transient. This means, I suggest, that waves of technological innovation work together to multiply the temporal frameworks in which we live, making us both 'young' and 'old' in the same instance. From this perspective, becoming older is an experience that relates directly to involvement and participation in sociotechnical assemblages. This experience, however, is characterised by complexity and multiplicity.

Taking her own experience of being 'locked out' of the dominant chains of food production and distribution – the fast food industry – Leigh Star (1991) suggested that in trying to understand the relationship between science, technology and identity, social scientists should not focus solely on the entrepreneur, the inventor, the 'man of ideas' to whom change is credited in Schumpeter's theory of innovation (and in early versions of Actor-Network Theory). Instead, we should focus on those at the margins of sociotechnical assemblages. This will be an appealing proposal for those concerned with older people in society, exclusion and isolation being a widespread experience shared by older individuals. However, Leigh Star reminds us that exclusion from one network is associated with membership in another. This entails 'acknowledging the primacy of *multiple membership* in many worlds at once for each actor in a network' because only through this recognition is it possible to understand how '*multiple marginality* is a source not only of monstrosity and impurity, but of a power that at once resists violence and encompasses heterogeneity' (Star, 1991: 30, my emphasis; Bowker and Star, 1999; Lampland and Star, 2009: 4–9).

In outlining this programme of research, Leigh Star was combining the ideas of Actor-Network Theory and Haraway's concept of the Cyborg, discussed in the previous section, with the neglected tradition of research about 'marginal men' within the Chicago School of Sociology. One of its founders, Robert Park, drawing on Simmel, argued that the multiplication of memberships and associated selves was an essential characteristic of modernity, marginality being one of its defining experiences (Park, 1928). Rather than following subsequent sociological research that emphasised 'role strain' of marginalised others against the need to maintain a unitary self, Leigh Star takes Park's proposition as a point of departure for research on social aspects of science and technology. Instead of taking the 'march of technology' – standardisation – as one singular movement shaping identities and ways of life, Leigh Star challenges us to instead dwell analytically on multiplicity.

This proposal is a key source of inspiration for the approach I am putting forward in this book. As I have argued in the previous section, opening an enquiry into the relationship between science, technology and ageing requires a commitment of diversity and multiplicity. This was, in turn, underpinned by wanting to investigate the *agentic* possibilities of ageing as an issue. I rejected the implication that agency equates straightforwardly with restoring 'voice' and ascertaining the ontological value of 'experience'. I suggested that there are problematic assumptions about the role of social science in restoring agency in a context of increased calls for 'activity and engagement' of older people. This is not to say however that there is no value in exploring and understanding individual experience, but its significance is heightened when it is seen, as Leigh Star suggests, in a relational lens, as an enactment of a particular form of subjectivity or self, mediated by particular forms of knowledge and artefacts – within particular agencements. Indeed, the value of experience, as Star's reading of Park contends, might be that it provides a model of the complexity and uncertainty of the agencement that produced it (Moreira, 2012b).

In contrast to the previous approaches explored in this section, this book aims to bring the relationship between science, technology and ageing to the fore of the research agenda. To do this, it can neither 'black-box' technological innovation nor assume the existence of one, homogeneous technology-related ageing experience (e.g. exclusion from technological networks). If both technoscience and ageing are opened up for investigation, we require a different trope, a figure guiding our conceptual and methodological imagination, without which we risk reverting to the comfort zone of negative, fatalistic narratives of technology that oppose standardisation to lived experience. This is where I think Leigh Star's proposal can be of use. By drawing our attention to the fact that marginality and membership are multiple, Leigh Star's work inspires us to explore the simultaneous plurality of sociotechnical worlds.

Indeed, it was responding to Star's and others' (e.g. Lee and Brown, 1994; Haraway, 1997) critique of Actor-Network Theory's insensitivity to difference and multiplicity within and without sociotechnical assemblages, that Law and Mol outlined their approach to STS. Their proposal is that it is possible to study multiplicity with the basic conceptual tools of Actor-Network Theory, but only if we don't overstate the centrality of its key terms, actor and network (Mol and Law, 1994; Latour, 1997). Instead, researching sociotechnical multiplicity entails 'de-centring' our analytical models and causal stories (Law, 2002, 2004, 2008, 2015). Instead of neatly arranged arrow or regular waves, what we might end up with are patchwork stories. As Law and Mol put it,

> materials and social – and stories too – are like bits of cloth that have been sewn together [...] there are many ways of sewing [and] there are many kinds of thread. [To follow this trope] it's to attend to the specifics of the sewing and the thread. It's to attend to the local links. And it's to remember that a heap of pieces of cloth can be turned into a whole variety of patchworks.
>
> (Law and Mol, 1995: 290)

Drawing upon Strathern's (2005[1991]) notion of 'partial connections', Law and Mol choose the feminist metaphor of the 'patchwork' to imagine the complexity of sociotechnical processes (see also Law and Mol, 2002).The trope of the patchwork evokes a tension between unity and multiplicity, of unrelated patterns, or fractional coherence. It belongs to a family of tropes and figures such as the manifold, the composite, the 'fractiverse' that attempt to reimagine complexity as crucially undermining social science's capacity to tame uncertainty (see Chapter 3). These concepts are underpinned by the argument that if complex systems are characterised by unpredictability, the social science approach to these can only be 'based on the dynamics of non-equilibria, with its emphasis on multiple future, [...] intrinsic and inherent uncertainty' (Wallerstein, 1996: 61). This in turn leads to the realisation that understanding and mapping the 'otherwise' is fundamental in social sciences' engagement with complexity (Law and Mol, 2002). From this perspective, complexity arises from the fact that objects are both singular and multiple; they comprise a multitude of different realities which, through the work of making connections, partially overlap (Moreira, 2000, 2006). Research on the dynamics of sociotechnical assemblages becomes then the business of following the threads within and between pieces of cloth of the patchwork (Moreira, 2004).

But multiplying our stories also has another purpose. As we saw above, sociotechnical worlds embed particular distributions of agency, power and recognition. Further, the stories we tell about how those worlds came about and relate to each other are likely to partake in those narratives themselves (Moreira, 2000; Law, 2015). So, for example, Schumpeterian economics embodies a heroic version of the entrepreneur that often buttresses institutionalised forms of power and control in the 'knowledge economy'. Similarly, gerontological discourses on the 'ageing body' shape professional interaction with older people (Katz, 1996). It would be unwise to think that social science theories and narratives, even those of critical gerontology (Lassen and Moreira, 2014) are any different in this regard. The patchwork trope leads us to ask the question: 'what would happen if we abandoned the idea that [stories] describe a single set of [realities] and instead [take up] the idea that they are performing different distributions?' (Law, 2002: 158). The aim of this book then is to take up this idea to explore the relationship between science, technology and ageing.

Chapter 3

Assembling the 'ageing society'

In the previous chapter, I suggested that to address the challenges of the ageing society, we should focus on developing a conceptual framework that tackles the uncertainty, multiplicity and complexity of the scientific practices and technological processes that sustain it, and that this was best done through an engagement with the science and technology studies scholarly corpus. In particular, I argued that both sides of the nexus of ageing and technoscience should be open to investigation, resulting in a programme of research concerned with understanding how distributions of agency are enacted and pragmatically relate to each other. This, in turn, was motivated by the aim to genealogically trace the 'the heterogeneity of what was imagined consistent with itself' (Foucault, 1991: 82): the ageing society. But what is the ageing society, from this knowledge-related, epistemic perspective?

In the introduction, I suggested that the usual answer to this question relates to the social, economic and political consequences that are seen to emerge from changes in population structure. Experts and policy makers disagree about what those consequences are, some viewing them as dictating fundamental changes to the nature of citizenship, while others see them as an opportunity, fostering technological and social innovation and creativity. I have further argued that rather than seeking to assess the rightfulness of each position there is work to be done in understanding the forms of knowledge, the institutional basis and the normative foundations of each of the arguments. This means that instead of asking for the facts that make the ageing society – its population dynamics, etc. – we ask: how, through what means and for what reasons did our societies come to be seen and described as 'ageing societies'? How did the label of 'ageing society' become an authoritative and accepted description of our shared lives, of our present condition and common future? How did it come to be 'gathered in a thing', and what technoscientific practices and processes 'make it exist and maintain its existence' (Latour, 2004: 246)? This is the question this chapter, drawing on the approach previously outlined, aims to address.

Taking as a point of departure key characterisations of the 'ageing society' articulated since the turn of the century, I then outline a genealogy of the relationship between the 'population problem' and the economy, articulated explicitly in actuarial sciences in the nineteenth century, and ultimately in the emergence of state-backed social security in liberal democracies. However, I suggest that the ageing society is in fact a questioning of the epistemic infrastructure of the liberal welfare state, a problematisation especially evident in the 'social security crises' of the early 1980s. As I will demonstrate, those crises were principally about the validity and reliability of expert knowledge, in that they concerned the failure of demographic calculations and economic forecasting to accurately predict increased life expectancies and correlated balances of social security programmes. In the last section of the chapter, I suggest that this recognition of the epistemic but unstable underpinning of the 'ageing society' served as the context to rebuild a research agenda for the 'ageing society' from the 1980s to the present day.

Ageing society at the turn of the twenty-first century

Since the turn of the century, the 'ageing society' has come of age. This transition was marked by a variety of national and global organisations identifying population ageing as one of the main challenges faced by humanity, along with climate change. The tone was perhaps set by the United Nations 'International Year of Older Persons' in 1999, which, under the theme of the age-inclusive society, marked a shift in ageing policies. Policy makers declared that whereas previous programmes were 'focusing upon a static age group apart from the rest of the population, efforts [were then being] made to view ageing as merely one stage in the course of individuals' lives' (Kalache and Kickbusch, 1997: 4). This meant that programmes should emphasise the commonalities and interdependencies between older people and the rest of society rather than single out a specific group. In that same year, the European Commission organised a series of policy conferences on 'active ageing' that also proposed an abandonment of the idea of 'the elderly' as a fixed group (Avramov and Maskova, 2004). This political orientation became later crystallised in a comprehensive approach to ageing through a coordination of 'mutually reinforcing policies' to tackle 'the economic, employment and social implications of ageing' (Lisbon Council, 2000: 5). Similar pronouncements, in the same context, were made at the 2005 Whitehouse Conference on Ageing and in the United Kingdom by the Science and Technology Committee of the House of Lords, following their major review of scientific evidence on ageing (House of Lords, 2005). Ageing was no longer a concern for older people alone.

Indeed, in these policy statements, it is not difficult to identify an emerging consensus about the need to depart from the idea that the 'problems of ageing' belonged solely to the elderly. While acknowledging the significance of the challenge brought to bear by demographic ageing, policy makers and experts appeared to agree that instead ageing and the 'ageing society' were to be seen as shared, societal problems. As the Secretary-General of the United Nations, Ban Ki-moon, put it in a 2012 report published by the UN Population Fund and HelpAge International,

> the social and economic implications of [ageing] are profound, extending far beyond the individual older person and the immediate family, touching broader society and the global community in unprecedented ways.
>
> (UNPF/HAI, 2012: i)

With this assessment, the UN Secretary-General was advancing the idea that while ageing and its consequences are most visibly and directly experienced by older individuals and their close social networks – family, friends, etc. – there are indirect, extensive – but mostly invisible – implications for the wider society, and for what he labelled the 'global community'. There are two key dimensions to this statement.

One relates to the scale of the 'community' that is seen to be affected by demographic ageing. It is not only that the consequences of ageing affect society as a whole, but also that this society is now conceived as a global entity. According to the same report, although mostly associated with the developed world, demographic ageing is occurring fastest in the developing world. This process, it is argued, further reinforces social, economic and technological inequalities between the Global North and South, as developing countries are growing old before they become rich (Kalache *et al.*, 2005). Linking the 'ageing society' with the dynamics of the global economy also problematised the ways in which demographic ageing had been framed and managed within national boundaries, and motivated the need to 'emphasise the power of knowledge as a policy tool' (Olshansky *et al.*, 2011). This is linked to the second dimension of the UN statement.

In a context of global ageing, policy makers argued, it was necessary to create the right epistemic infrastructure to make the implications of the ageing society visible, tangible and actionable. Thus, in Ban Ki-moon's statement, the link between the individual case and the 'wider society' is tellingly described as 'touching', to connote the fragile, almost imperceptible connection between the two. The difficulty with understanding the 'ageing society' was then more related to the fact that it was not available to immediate experience, to the 'naked eye', so to speak. To realise the link between individual experience of ageing and its wider, aggregate consequences, it was necessary to draw on a particular type of lens, a particular way of looking at and knowing the ageing process. What are the ways of looking, the knowledge practices and institutions that are so fundamental for the ageing society to be 'gathered as a thing'?

One way to answer this question is to look at how the consequences of population ageing are outlined. To do this, it is perhaps instructive to draw on another report published in that same year by Eurostat – an organ of the European Commission which provides statistical information to the institutions of the European

Union. The report summarises the wider societal implications of population ageing as resulting in,

- pressure on public budgets and fiscal systems;
- strains on pension and social security systems;
- adjusting the economy and in particular workplaces to an ageing labour force;
- possible labour market shortages as the number of working age persons decreases;
- the likely need for increased numbers of trained healthcare professionals;
- higher demand for healthcare services and long-term (institutionalised) care;
- potential conflict between generations over the distribution of resources.

(Eurostat, 2012: 7)

Mostly dealing with what is commonly known as 'facts and figures', the report thus neatly identifies the knowledge domains pertaining to the ageing society: the knowledge of population dynamics (demography), public finances, labour market economics, health and social care and the socio-economic relations between age groups. One of the most striking aspects of this list is the weight of the economic aspects of the ageing society: resources, demand, labour force, markets, budgets, fiscal systems. The report proposes that there is a relationship between the factual reality of population ageing, of the increased proportion of older people in the European population, and the hard reality of economics, of money, markets and budgets. This bond is further evinced in the following graph (Figure 3.1).

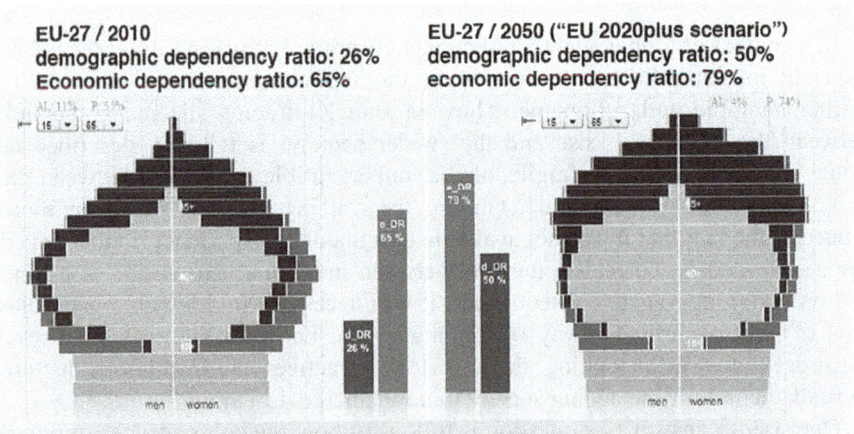

Figure 3.1 Fritz von Nordheim.

Source: reproduced from Lassen and Moreira, 2014.

Taken from a presentation by Fritz von Nordheim at a European Active Ageing Conference in December 2012, the graph represents the current and future age pyramid for the 27 countries of the European Union. The age pyramid is a commonly used bilateral histogram that divides the population into gender and age cohorts (see Chapter 4). Its main advantage in comparison with tabular distributions of age groups is that, by its shape alone, it provides the viewer with an assessment of the age structure 'at a glance'. Usually, pyramids with larger bases are seen to represent growing, 'younger populations' while constricted pyramids are seen to belong to populations in decline.

In the example provided, to this demographic assessment is added an economic qualification of the population, whereby in each age cohort are nested calculations of the proportions of the active and inactive members of each age group. In the graph, Nordheim, a social scientist specialising in European welfare systems, known for his policy work in the Employment and Social Affairs domain, brings the two dimensions of the 'ageing society' together. On the one hand, looking at the outlines of the two pyramids, we are able to observe that the European population is expected to age considerably in the next four decades, as the older age groups become more prominent. On the other hand, looking within the age group bars, we become aware of distributions of working capacity within the population between those actively working (in light grey), those unemployed but capable of working (grey) and those incapable of working (black). From this we quickly glean a relative increase of the proportion of the population not participating in the labour force. By classifying different age groups in this way, the graph brings to the fore the notion that links demographic ageing and economics: that an ageing population will see a decrease in the number and proportion of active, productive members of a society. This is seen to affect not only the economic efficiency of a nation, but also the distribution of resources across generations, with a smaller proportion of younger, more active members holding more of national income. With current arrangements relating to pensions, social security and the financing of health and social care, the relative gain in income becomes lost to intergenerational transfers.

One way of tracing the evolution of these transfers as an aggregate is known as the dependency ratio, which calculates the relationship between active members of the population and those receiving any transfer in the form of education, welfare, pensions, etc. While the most widely used dependency ratio includes both individuals below and above 'working age' (15–16 and 65 years), in the context of demographic ageing, experts tend to use the more specific 'old age dependency ratio', which calculates the number of persons aged 65 or more as a proportion of those of working age. On the basis of these premises, Eurostat calculates that in 2010 there were almost four persons of working age for every older person (26 per cent), and projects that in 2060 this ratio will be closer to 2 : 1 (50 per cent) (Eurostat, 2012: 32).

Nordheim's graph above builds on these figures to calculate an even more detailed ratio between population not in the labour force and those actively working: the economic dependency ratio. This more specific ratio attempts to shift the focus from chronological age to the actual working population, regardless of age (see Chapter 5). Curiously, however, instead of projecting a dramatic increase in the dependency ratio in 2050, as Eurostat proposes for the old age-related calculation, the economic dependency ratio is only expected to rise about a quarter in the next four decades. This is because Nordheim's projection is based on the realisation of adequate pension reforms in the EU member states prior to 2020 – an imagined possibility known as the 'EU 2020plus scenario' – whereby there is a consistent postponement of retirement mirroring life expectancy, health and functionality gains in the population (see Chapter 7).

Knowing the series of assumptions, notions and expectations on which the graph relies makes it acquire a different set of meanings. While previously we might have engaged with the graph as an accurate depiction of the evolution of the relationship between demographic ageing and economics in the European Union, now it represents also, perhaps mostly, a proposition in economic policy. What we now see is a particular solution to a problem, a pathway within a number of possible choices. What we now become aware of is that those other possibilities are not available for us to consider. For most of us, this moment of doubt would not last long, returning promptly to the issue at stake and engaging helpfully in trying to support a solution. But what happens if, as this book proposes, we open this moment of doubt and explore the uncertainties within it, if we attempt to dwell in this space of disbelief for long enough to think about the knowledge assumptions of the 'ageing society'. How did we come to frame it almost exclusively by the relationship between demographic ageing and its economic implications? How, of all the possible implications of ageing, did productivity (or 'economic activity'), budgets and transfers come to define the quality of the social relationships that characterise the 'ageing society'?

Outline of a genealogy of the 'ageing society'

To understand the question posed in the previous section, it is necessary, I suggest, to grapple with the key element of the 'ageing society': its population dynamics. In other words, the ageing society is above all a 'population problem'. In his theorisation of governmentality, Foucault describes a contrast between two regimes of power-knowledge and forms of security. In sovereign-type states, territorial authority emanates directly from a central ruler, deployed through strategic knowledge about 'enemies'. In liberal, governmentalised states, power is reliant instead on knowledge and management of populations. Defined as 'a global mass affected by overall pressures of birth, death, production, illness' (Foucault, 2003: 243), the population is understood through a series of 'dispositifs' focusing on these aspects of the 'mechanics of life'. In particular, it is contingent on the application of methods of data formatting and procedures of statistical analysis that render such mechanisms visible. Of significant relevance, in view of how the implications of the ageing society are imagined by key policy institutions such as the UN, is Foucault's suggestion that these mechanisms are 'aleatory and unpredictable when taken in themselves or individually, but which, at the collective level, display constants that are easy, or at least possible, to establish' (Foucault, 2003: 246). Issues such as the 'ageing society' can only exist at the aggregate, population level.

Also important in Foucault's conceptualisation of what he labelled the 'bio-politics of security' was the concern with the 'economic'. Statistical measurement and calculations were in this respect considered part of the very apparatus that enabled economic freedom. This is because the governmental regime's bio-political technologies were geared towards deploying circulation of the contingent 'natural flows' of economic activity, through incentives and deterrents. In this respect, the 'invisible hand' of the market works to limit and shape the power of administrative authorities, in that it questions, through its 'court of veridiction' (prices, etc.), the capacity of government to control individual action (Foucault, 2010: 30). It is from this perspective that it is possible to appreciate Foucault's now famous definition of governmentality as the 'conduct of conduct'. These are forms of knowledge-mediated administration that rely on the provision of implements to guide the comportment of individuals, relying thus on individuals' management of their own conduct through the use of 'technologies of the self'. For Foucault, these are productive processes, whereby the capacities of rational autonomy and self-determination – the *homo economicus* – are generated in entanglement with enactments of liberal, soft power.

While Foucault has been criticised for his lack of attention to the specificities of statistical reasoning (e.g. Desrosières, 2008: 101–117), and differences within liberal political economies, his work on biopower provides a useful handle to explore the conjunction between the management of populations and the economy. But to fully realise this we need, as suggested by Tellman (2013) and Dean (2015), to recast the role of the Malthusian 'problem of population' in the consolidation of biopower. In *An essay on the principle of population*, Malthus (1872[1792]) argued that increases in population would be limited by the more limited capacity to increase food production, or scarcity. Scarcity, he argued, was the key regulatory mechanism in population dynamics, directly shaping the curve of mortality ('negative checks') through famine and disease, unless 'positive checks' of sexual abstinence and/or abortion were implemented to control population growth. While Foucault focuses on Adam Smith and the *physiocrats* to understand the role of the market and wealth in a liberal political economy, Malthus' work placed, particularly through Ricardo's work on rent, scarcity at the heart of modern economic reasoning. Such reasoning pervades the listing of the consequences of the 'ageing society' provided by the Eurostat, with its references to 'pressures', 'strains', 'shortages' and 'conflict' over resources. As Dean argues, in attending to the Malthusian 'population problem', it is possible to understand how 'the laws of life are the condition of possibility of economic knowledge' (Dean, 2015: 22).

It is somewhat surprising that Foucault linked scarcity with disciplinary power rather than biopolitical power, given his interest in the development of actuarial science. In his view, 'insurance or regularizing technology' was central to the apparatus of governmentality (Foucault, 2003: 222), the notion of population being founded on the epistemological assumption that it possessed its own regularities, 'its number of dead, its number of sick, its regularity of accidents' (Foucault, 2009: 107). For Foucault, governmentality was underpinned by 'the absolutely capital concept [...] of risk' (Foucault, 2009: 87; Ewald, 1991). As is well recognised, one of the central techniques of capturing these regularities and risks is the life table. This is a tabulation that shows, for each age, the probability of death within a year. It is a fundamental tool both in classical actuarial science, enabling the calculation of insurance rates, and in demography, to determine age-specific life expectancy – and, through this, to make predictions of future age structures, such as the EU27/2050 (see Figure 3.1).

The importance of the Malthusian 'population problem' in both actuarial science and demography is embodied in Gompertz' Law of Mortality. Gompertz, a practising actuary working for the Select Parliamentary Committees on Friendly Societies in the 1820s, proposed that mortality rates increase in a logarithmic progression (Gompertz, 1825; Kirkwood, 2015). Gompertz' law is interesting because he decomposed death into two categories: those caused by chance, which would act in a constant, linear manner; and those caused by 'deterioration, or an increased inability to withstand destruction' (Gompertz, 1825: 517). Chance death could be seen as representing a normalising transformation of the arithmetic rate governing food production in Malthus' model, in that it replaced the catastrophic 'checks' on population growth by famine and disease (Dean, 2015) by a smooth, unceasing force. Although chance death was excluded in Gompertz' equation – later reintegrated by Mackeham – his technique enabled a calculation of 'life contingencies' that quickly became part of the standard toolkit used by insurance companies. This practice routinely attributed a monetary value – *a worth* – to a specific age.

While the explicit valuation of life was not new in actuarial science, it was bolstered by political legislation to regulate Friendly Societies' role in providing life, burial and sickness insurance for the working class (O'Malley, 2002, 2009), which by 1815 covered about one third of English households (Thane, 2000: 194). These reforms, much swayed by Gompertz' work for the Parliamentary Committee (above), required that credentialised actuaries be responsible for the setting of annuities in the Societies. This embedding of actuarial expertise on the cooperative insurance sector aimed to create the tools for incentivising thrift in the emerging industrial labourer population. In this respect, Gompertz' division between chance death and deterioration formalised in actuarial classification the progression from 'savage life' – ensnared by the accidental – to 'civilised life' – characterised by planning and restraint – that is at the heart of Malthus' theory of population (Malthus, 1872[1792]; Tellman, 2013; Dean, 2015). Indeed, in Gompertz' original paper, accidental and unavoidable death coexisted in his analysis of existing life tables (Hooker, 1965), but the latter modelled the unadulterated process of ageing and death. Given Gompertz' admiration for Newton, it can be

TABLE V.—Logarithms of the accommodated chances of living 10 years, deduced from the value of an annuity for 10 years, at 5 per cent. from the actual tables of mortality, and considered equal to a geometrical series of ten terms, of which the common ratio is the same as the first term, and the tenth term the accommodated chance; and to find the accommodated chance for 5, 7 years, &c. without a table calculated for the purpose, it may be considered sufficient to multiply by ,5 ; ,7, &c. the accommodated ratio in this table when extreme accuracy be not required.

Age.	Carlisle.	Deparcieux.	Northampton.	Age.	Carlisle.	Deparcieux.	Northampton.
0	1,6892	—	—	52	1,9172	1,9006	1,8523
1	1,6763	—	1,7044	53	1,9098	1,8957	1,8471
2	1,8699	—	1,8356	54	1,9013	1,8901	1,8417
3	1,9159	1,9166	1,8790	55	1,8915	1,8853	1,8357
4	1,9401	1,9315	1,9081	56	1,8803	1,8799	1,8294
5	1,9586	1,9411	1,9220	57	1,8680	1,8732	1,8228
6	1,9686	1,9486	1,9369	58	1,8513	1,8673	1,8156
7	1,9737	1,9544	1,9476	59	1,8435	1,8601	1,8081
8	1,9764	1,9592	1,9550	60	1,8318	1,8511	1,7998
9	1,9773	1,9637	1,9586	61	1,8243	1,8398	1,7908
10	1,9768	1,9669	1,9592	62	1,8171	1,8264	1,7811
11	1,9754	1,9679	1,9582	63	1,8090	1,8120	1,7699
12	1,9742	1,9669	1,9566	64	1,7974	1,7946	1,7576
13	1,9729	1,9658	1,9546	65	1,7860	1,7735	1,7431
14	1,9716	1,9704	1,9521	66	1,7703	1,7510	1,7267
15	1,9704	1,9628	1,9490	67	1,7506	1,7270	1,7083
16	1,9698	1,9609	1,9455	68	1,7107	1,7017	1,6879
17	1,9694	1,9600	1,9419	69	1,7005	1,6754	1,6651
18	1,9693	1,9586	1,9388	70	1,6689	1,6480	1,6402
19	1,9690	1,9574	1,9358	71	1,6319	1,6167	1,6126
20	1,9685	1,9559	1,9337	72	1,5936	1,5841	1,5823
21	1,9679	1,9554	1,9321	73	1,5563	1,5500	1,5487
22	1,9670	1,9549	1,9311	74	1,5269	1,5119	1,5117
23	1,9659	1,9544	1,9298	75	1,4940	1,4711	1,4723
24	1,9644	1,9540	1,9289	76	1,4642	1,4218	1,4308
25	1,9628	1,9534	1,9277	77	1,4344	1,3684	1,3846
26	1,9573	1,9531	1,9264	78	1,4007	1,3134	1,3307
27	1,9591	1,9524	1,9257	79	1,3538	1,2497	1,2644
28	1,9570	1,9521	1,9238	80	1,3134	1,1876	1,1900
29	1,9556	1,9518	1,9226	81	1,2582	1,1214	1,1101
30	1,9552	1,9514	1,9211	82	1,2043	1,0609	1,0234
31	1,9548	1,9514	1,9196	83	1,1765	2,9688	2,9341
32	1,9540	1,9514	1,9180	84	1,0727	2,8536	2,8592
33	1,9528	1,9515	1,9164	85	2,9939	2,7199	2,7813
34	1,9513	1,9517	1,9146	86	2,9166	2,5736	2,7003
35	1,9485	1,9522	1,9126	87	2,8490	2,4254	2,6149
36	1,9477	1,9528	1,9104	88	2,8055	2,1943	2,5369
37	1,9452	1,9534	1,9083	89	2,7537	3,9129	2,4179
38	1,9437	1,9527	1,8057	90	2,6695	3,5265	2,2414
39	1,9406	1,9517	1,9031	91	2,6658	3,0266	3,9356
40	1,9383	1,9506	1,9001	92	2,7323	4,3694	3,5037
41	1,9372	1,9488	1,8973	93	2,8031	5,2971	4,7375
42	1,9365	1,9466	1,8943	94	2,8355	—	5,5769
43	1,9365	1,9438	1,8915	95	2,8107	—	7,0496
44	1,9366	1,9403	1,8882	96	2,8279	—	—
45	1,9367	1,9361	1,8848	97	2,7589	—	—
46	1,9366	1,9308	1,8810	98	2,6695	—	—
47	1,9358	1,9263	1,8767	99	2,5111	—	—
48	1,9351	1,9200	1,8740	100	2,1629	—	—
49	1,9328	1,9158	1,8631	101	3,5689	—	—
50	1,9292	1,9098	1,8621	102	4,3245	—	—
51	1,9233	1,9027	1,8571	103	6,0595	—	—

Figure 3.2 Gompertz calculations of 10 year survival chance based on Carlisle, Deparcieux and Northampton life tables (in Gompertz, 1825).

said that approximation to this ideal was not only closer to natural truth but also to ideal moral worth.

Efforts to professionalise the Friendly Societies' insurance schemes continued through the nineteenth century. They were particularly boosted by the introduction of the first old age pension programme by Bismarck in 1889, which campaigners contrasted with the inadequacies of the Poor Law and Friendly Societies in providing for older people in Britain (Thane, 2000: 198–199). These failures were brought to light by various surveys of poverty in old age presented to a succession of parliamentary committees on the issue, such as Booth's study of labourers in London. It was uncertain however whether this was due to existing schemes or a lack of planning and restraint by the poor. Proposals for a state-managed old age pension were thus hampered by concerns over its role in discouraging 'providence' among the working classes, and, among other things, by the Friendly Societies' mistrust of a centralised scheme. There was also debate about whether this should become a contributory scheme, as those provided by Friendly Societies, or just reliant on means testing. When eventually the Old Age Pensions Bill came to parliament in 1908, qualification hinged on a test of demonstrable willingness to work – the 'industry test' – and, importantly, an age limit.

In combining these two qualifying criteria in the administrative management of old age pensions, the British scheme crystallised the two key categories that much later were still to underpin the construction of the EU27 age pyramids analysed in the previous section: activity and age. In this respect, the scheme represents a paradigmatic example of the liberal formatting of the relationship between work and age, also present in the French or German social insurance approach. Crucially, it made actuarial expertise, in calculating present and future life expectancies, essential for the deployment of government budgets. This supports, in relation to old age pensions, Ewald's (1991) claim that the emergence of liberal welfare states marks their becoming 'insurance societies', in which new social bonds are enacted by insurance underwritten by the state. As Beveridge put it in his report on social insurance after a visit to Germany in 1908, such a 'scheme sets up the state as a comprehensive organism to which the individual belongs' (Beveridge in Hennock, 1987: 136) The use of 'forecasts, statistical estimates and overall measures' of ageing became thus linked to figures on economic growth, employment, taxation income, etc., in the machinery of the state, a tie that would only become more durable with the consolidation of the welfare state in the next four decades. Ageing became part of what we, after Scott (1998), could label the 'state gaze' on society.

Reinventing the ageing society

A genealogical exploration of the ageing society clarified how it belongs to the set of population problems enacted by liberal biopolitical regimes. In this process, ageing became a key articulation of the relationship between mechanisms of life and the economy, embodied in the practices of actuarial science. As these were incorporated in the administration of populations by the implementation of old age pensions, it made the collection of data about age, and 'old age' in particular, a key activity of emerging welfare states. But these components are not enough to elucidate fully how the 'ageing society' 'gathered as a thing' a few decades later. In the first section, we suggested that the ageing society also assembles when the underlying – actuarial – assumptions of programmes such as pension schemes and health and social care financing are challenged by the process of demographic ageing. The ageing society is, to put it simply, a difference between the actuarial forecast and the contemporaneous reality of the population. This, I will suggest in this section, is the result of the institutions of 'old age' being captured by another set of sociotechnical relations.

Although there are mentions of the 'ageing society' as far back as the 1930s, for reasons we will explore in the next chapter, it is only in the 1970s that the concept, as we know it today, became established. Key in this process was the publication of the Social Policy, Social Ethics, and the Aging Society Report, by the Committee on Human Development of the University of Chicago. Drawing on a research project that combined the expertise of demographers with that of economists and social policy scholars, the book attempted to examine the 'future of ageing and the ageing society' (Neugarten and Havighurst, 1976: iii). To do this, experts assessed the 'present pattern of populations' and crafted 'predictions' to derive consequences for the organisation of health and social services for the 'elderly' in the future. Their assessments and predictions were explicitly labelled as uncertain, as emerging out of novel, unexpected conditions for which it was necessary to create new forms of knowledge and expertise.

Indeed, almost a decade later, in what can be considered a direct offspring of the CHD research initiative, the US Committee on an Aging Society, could still state that,

> The ways in which an ageing society might be a different society, in other than demographic characteristics, are not entirely clear. But it is evident that the changing age distribution of the population will have major implications, at the very least for the following:
>
> - financing, development, organization, and use of health care systems;
> - patterns of family life, social relations, cultural institutions, living arrangements, and physical environments;
> - distribution of jobs among older and younger workers, as well as the earnings, status, and satisfaction that these jobs may provide, within the context of age discrimination laws, seniority practices, and technological innovation;
> - economic aspects of providing retirement income through various public and private mechanisms;

- quality of life of the population throughout the life course including functional status, well-being, legal status, and personal autonomy; and
- an ever-shifting agenda of related public policy issues.

(Committee on an Aging Society,1985: ii)

One of the striking aspects of this list is its similarity to the EU description of the consequences of the ageing society three decades later, on the other side of the Atlantic. Indeed, the US Committee can be said to have laid the foundations of policy framing of the ageing society. These foundations relied, as one would expect, on a historically amassed collection of forms of knowledge about the ageing process brought to bear by the advent of old age pensions in the liberal welfare state. In the US, this had been instigated through the interest of the Macy Foundation in understanding the implications of the Social Security Act of 1935. This in turn had been activated by Edmund Vincent Cowdry's (then a respected cell biologist) concern about how, in the Great Depression, such programmes sparked potential tensions between the functional capacities of older people and bureaucratic standards (Park, 2016: Chapter 2: see also Chapters 5, 6 and 9). Brought together by the Macy Foundation, Cowdry and Lawrence Frank (see Chapter 2) were responsible for organising what was to become the bedrock of gerontology as a field of knowledge: the Woods Hole conference on the 'problems of ageing' (Cowdry, 1939; Katz, 1996: 77–103). In particular, it had created a model of social organisation of knowledge – more recently known as multidisciplinary – where different strands of expertise (biology, philosophy, demography, etc.) come together to tackle a shared problem.

While the implementation of this ideal was fraught with difficulties in the next decades (Achenbaum, 1995), it is undeniable that it served to frame the composition of the US Committee on an Aging Society, which included physician-scientists – the Nobel Prize-winning virologist and paediatrician, Frederik C. Robbins – economists, actuaries, epidemiologists, social scientists but also architects, occupational therapists and lawyers. The remit of the Committee was to collect, review and assess the 'state of knowledge' relating to specific areas of concern, such as the productivity of older people, health and the built environment. In so doing, they gathered together the various programmes of ageing research that had been sparked by the Woods Hole conference to examine the present and future of the ageing society. From this perspective, it is significant that the Committee's assessment of this corpus of knowledge is saturated with uncertainty, the nature of the 'ageing society' itself not 'being entirely clear'.

Crucially, this uncertainty relates to the core body of evidence on which the category of the ageing society is built: demography. In their paper reviewing 'demographic trends and projections', Serow and Sly suggest that,

> it is important for us to know much more about patterns of mortality and longevity and how these are changing. This may sound simple, but the

complexity of such knowledge is evident when we consider that the size of this population is influenced by historical patterns of fertility and mortality to the point at which persons enter old age, as well as by patterns of mortality throughout the older years of life – to say nothing about the patterns of immigration and emigration over the whole course of life.

(Serow and Sly, 1988: 43)

Casting research on the demography of ageing as 'relatively recent vintage', Serow and Sly provide an overview of the complex set of variables that enable a calculation of the changing relative size of 'old age' cohorts. Added to this complexity is the sense that the quality of the data and of the techniques used to compute demographic forecasts is uncertain, questioning whether 'the knowledge provided by the demography of ageing can help policymakers' (Serow and Sly, 1988: 43). Such a precautious assessment of the capacity of demography to produce accurate and reliable forecasts contrasts with the epistemic authority it emanated within earlier welfare 'insurance societies'. What we encounter instead is a situation where it is 'important to know much more' than previously thought.

The lack of confidence in demographic forecasts is less surprising if we consider that Serow and Sly were writing just a few years after the 'social security crisis' of the turn of the 1980s. This event had publicly questioned the actuarial assumptions upon which old age pensions and other programmes had been established. Kenneth Manton, a prominent US demographer who was a central figure in the public debates on this 'social security crisis', recalls the processes as such:

Before the implications of U.S. population aging became nationally visible in the Social Security 'crisis' of 1982 and 1983, there was little analysis of the impact of the growth of the elderly and oldest-old population on national policy. [...] By 1982–1983 it was recognized that increased cohort size and improvements in life expectancy had major implications for Social Security and Medicare. Congressional hearings were held to examine the fiscal implications of fertility and mortality trends and to determine why the trends had not been identified earlier [...]. The systematic under-projection of the elderly population led the U.S. Senate Finance Committee (1983) to conclude that the official projections had received inadequate scientific input and the uncertainty of projections had been underestimated.

(Manton, 1991: 310–312)

As Manton's narrative suggests, the crisis had arisen from the political realisation of the implications of increased life expectancy for pension provision and health care support for older citizens. Both these systems had been built on the assumption that average life expectancy at 65 years of age would remain more or less constant. There were good reasons for this. From the 1930s until the 1960s, the Medicare decade, a consensus had formed within the ranks of population scientists and biologists that, as the National Center for Health Statistics would put it, 'further decreases [in death rate] as experienced in the past cannot be anticipated' (NCHS in Manton, 1991: 311). Such a prediction was underpinned by the view that increases in life expectancy had been driven by changes in child health and survival at early ages, but that these gains in life expectancy would not be reflected in the size of the older cohort because of the combination of the arch of development and old age, and a natural 'limit' in human longevity.

Breaking this consensus entailed re-examining the relationship between mortality and morbidity – the quantity and quality of life. In 1980, Fries had strongly reinforced the mainstream view by proposing that gains in the prevention of illness would lead to a 'compression of morbidity' against a limit of longevity (Fries, 1980; Fries and Crapo, 1981). This would approximate the curve of mortality to the 'ideal' proposed by Gompertz (above). It also promised a world where, by preventative medicine, time spent living with chronic disease was much reduced. Manton himself had been pivotal in questioning this scenario by arguing that while morbidity might increase in older cohorts, its severity on average would be reduced, leading to gains in life expectancy (Manton, 1982). This was a possible situation where more people were living for longer in lighter states of disability, with obvious implications for planning of social security and health care programmes. Similarly, Olshansky and Ault suggested, a few years later, that the onset of chronic diseases would, through a combination of health technologies and public health, shift to older age groups (Olshansky and Ault, 1986). Finally, there was the prospect that the effects of medical technology had done nothing more than to extend the number of years in which older people experienced disability (Schneider and Brody, 1983).

In this respect, the controversy was compounding the sense of uncertainty that emerged from the public realisation that 'official projections had received inadequate scientific input'. It was not only that demographic and actuarial calculations were deemed uncertain, it was also that experts could not agree on the instruments and measurements that should be used to evaluate the consequences of 'demographic ageing' on the public purse and the taxpayer. It is from this perspective that we can read Serow and Sly's irresolute review of the demographics of ageing. There were observable uncertainties not only about the knowledge base for decision-making but also about the institutional character of knowledge making procedures: was the administrative apparatus of the state adequate for collecting data related to such a complex, unstable object as ageing?; did actuaries, demographers and epidemiologists have the right tools to understand age-specific life expectancy or morbidity?; was there enough funding to understand the 'fixed' and malleable aspects of age-related illness?

A similar 'knowledge crisis' was also emerging in the United Kingdom. As in the United States with the Reagan administration, Britain had a few years earlier elected a government that wanted to promote economic prosperity through market-driven reforms and the restructuration of public services, including pensions. In this context, the Institute for Fiscal Studies (IFS), an independent think-tank concerned with the financial basis of state administration and social and economic policies, conducted a wide-ranging review of the social security system. Departing from the proposition that 'Social Security is "another British Failure"', Dilnot *et al.* (1984: 1) quickly identified the pension system as the major component of this debacle:

> Over the twenty year transition period [since the implementation of the Beveridge Report], costs [of social insurance] would rise by about 25 percent. This increase was wholly attributable to the rising burden of retirement pensions. The number of people of pensionable age was expected to increase from 5.6 million (12% of the population) in 1941 to 9.6 million [...]. In the event, there were 10.1 million people in this age group in 1971. [...] Although the number of pensioners was to increase markedly in the postwar period, the number of people of working age was actually expected to fall slightly as a result of the low birth rate between the two world wars. The consequence was that a fixed contribution was to be paid by a declining number of people while demands on the fund were rising.
>
> (Dilnot *et al.*, 1984: 12–13)

While just focusing on social security, and thus concerned with morbidity and chronic disease in the report, Dilnot and his colleagues could still link the problems experienced by the social security system to weakness in actuarial and demographic forecasts. In the British case, this was not only linked to the fact that the population of pensionable age grew proportionally more than expected; it was also that, because of this relative imbalance, those contributing to the scheme were fewer in number than would have been ideal. In other words, the IFS was suggesting that because of this statistical relationship between active and non-active populations, the system was not sustainable. This assessment is usually linked to cutbacks on the State Earnings-Related Pension Scheme (SERPS) by the Conservative government in 1986 (Macnicol, 2015: 56).

As should be apparent by now, we can take the various elements of these 'social security crises' as the model and the origin of the 'ageing society'. Seen from this perspective, the 'ageing society' is less about the societal consequences of population ageing, and instead should be viewed as a knowledge-related problem. The 'ageing society' stands for a concern with 'knowledge reflexivity', where there is increased, widespread awareness of the uncertainty of the knowledge base upon which key social and welfare institutions rely. In this respect, the 'ageing society' can be seen as part of a wider transformation in the institutional framing of knowledge processes in contemporary societies.

In sociology, these transformations are mainly associated with the emergence of what Beck (1986) labelled the 'risk society'. Beck, drawing on Luhman and Ewald, proposed that members of the 'risk society' experienced a paradoxical combination of increased public awareness of technological, man-made risks and a recognition of depending on the same forms of scientific knowledge to assess, and of technological innovation to solve, such risks. However, our analysis shows that, in the case of the 'ageing society', it is those technologies of risk assessment and control of contingencies that become problematised. The 'ageing society' appears at the confluence of failed past actuarial assumptions and the need to reimagine the implications of such failures in the future. In this regard, the 'ageing society' can be said to derive its very existence from the uncertainty of expertise and expert calculations.

For this reason, it is perhaps more useful to focus on this amplification of institutional emphasis on knowledge uncertainty. This is because what appears to characterise the 'ageing society' is not a crisis in public trust about population science or economics, but a process of collective questioning about, of opening the relationship between, knowledge making and social institutions, which puts uncertainty at the heart of the process. The 'ageing society' corresponds to the social, reflexive process which comes from the collective realisation that,

> [c]ontrary to what we might have thought [...], scientific and technological development has not brought greater certainty [but] in a way that might seem paradoxical, it has engendered more and more uncertainty and the feeling that our ignorance is more important than what we know. The resulting public controversies increase the visibility of these uncertainties. They underscore the extent of these uncertainties and their apparently irreducible character, thereby giving credit to the idea that they are difficult or even impossible to master.
>
> (Callon *et al.*, 2009: 18)

Such is the case of the 'ageing society' where controversies about life expectancy and disability in later life compound the already complex business of calculating budgets or transfers across generations. Importantly, the instability of actuarial and demographic forecasts brings into view the delicate foundations of welfare programmes, of the 'insurance society' as we described it in the last section. It opens those up for debate. The 'ageing society' labels the processes whereby members of a society engage in collective – often conflictual – negotiation about the nature of the link between knowledge and the politics of ageing, enacting and making visible how ageing-related expertise relates to a normative ideal of living together across the life course.

In proposing this, I am diverging from other understandings of the 'crisis of social security' of the 1980s and 1990s, that either see this problematisation as a result of an ideologically driven, neoliberal attack on the welfare state or suggest that knowledge uncertainty becomes a niche opportunity for the implementation of marketisation reforms. I do not deny the existence of such drivers in the reform of social security in the last three decades. However, if we depart from an epistemic understanding of the ageing society, the question should be instead focused on the knowledge making institutions that emerged within this 'age of uncertainty'.

Reimagining ageing

Recasting the crisis of the 'insurance society' as a crisis of knowledge provides a scaffold to explore the nexus between ageing and technoscience. If the 'insurance society' had relied on and reinforced models of ageing that enacted 'old age' as a biological, psychological and socio-economic stage in life (Katz, 1996), for which specific welfare programmes were required, the 'crisis of social security' was underpinned by opening up established knowledge on age-related functional decline, disability and life expectancy. As we will see in the remaining chapters of this book, this problematisation did not begin *de novo* in these years, but was instead a parallel formatting of the ageing process, which controversies on demographic forecasts poignantly linked to present and future ways of organising social relations. This connection posed the question: what kind of knowledge and knowledge making institutions do we require to manage the uncertainty that is inherent to the 'ageing society'?

This search for a new framework for knowledge on ageing was evident already in Manton's proposals to Congress in its hearing on Life Expectancy in 1983, for example. Recognising the irreducible uncertainty of actuarial calculations, Manton recommended the creation of two 'technical advisory groups', one concerned with the tools and technologies of forecasting and the other to 'evaluate epidemiological and biomedical evidence' of the impact of health technologies on mortality and morbidity patterns (Manton in Committee on Finance US Senate, 1983: 61). His proposals are important because they juxtapose the actuarial apparatus with a concern with biomedical and public health research. In imagining this new system of evidence gathering Manton was partially stabilising the list of entities that would be relevant for the 'ageing society' as pertaining to longevity, health and dis/ability.

Significantly, it is also within the context of the crisis of demographic forecasts that we see the establishment of the National Institute on Aging (NIA) in the United States, after many years of failed projects to do so (Lockett, 1983). Seen as a flagship research institution aiming to coordinate all nationally funded ageing research, the NIA had been set up to provide 'a comprehensive investigation of the normal, physiological changes with age' (Butler, 1977: 8). As we will see in Chapter 6, the idea of 'normal ageing' had been proposed as a conceptual device to imagine a distinction between avoidable and unavoidable morbidity and disability associated with ageing. Knowing and establishing such a 'standard' was seen to be a powerful alternative to the 'arbitrary' age limit for work, retirement and health care entitlement (Chapter 5). Indeed, as Robert Butler, the first Director of the NIA, put it in 1982:

> Economic perturbations have threatened the integrity of the Social Security System, [and have motivated] proposals to increase the age of social security eligibility. [...] Because of the increased age of the workforce and conflicts over retirement age [...], we must be able to assess properly the impact of aging on human performance.
>
> (Butler, 1982b: vi)

Butler explicitly describes the 'social security crisis' as a period of heightened uncertainty and controversy over retirement age and its relationship with function. His view is that the problematisation of age-graded programmes justifies and motivates a research programme on 'the impact of ageing on human performance'. Like Manton, Butler proposed that the key consequence of such debates is a focused research agenda on ageing and dis/ability, to establish what can legitimately be asked of and provided to older people. In so doing, he postpones the collective decision of raising the retirement age, and hinges such a choice on future technoscientific research and innovation, a future he hoped would be shaped by the activities of the NIA.

A similar type of imaginary supported attempts to reorganise ageing research in Europe. In France, where neoliberal models had less bearing, the emerging uncertainties about life expectancy and malleability of health and functionality in 'old age', most visible in the work of François Bourlière, justified the creation of the Unité de Recherches Gérontologiques within the Institut National de la Santé et de la Recherche Médicale (Lenoir, 1979). In Britain, questions about extended age-specific life expectancy and the role of biomedicine in this were also central to renewed discussions regarding the – ultimately failed – creation of a NIA-like institute in the United Kingdom (MRC archives).

But nowhere is this imaginary more visible than in the workings of the US Committee on an Aging Society, discussed in the previous section. Their symposia 'did not want to propose national policy but to explore areas of research that might contribute usefully to the weighting of national choices' (Committee on an Aging Society, 1985: vi). These areas were: health and ageing; ageing and productivity; and the role of assistive and 'environmental' technologies in later life. On health, the Committee, in alignment with Manton's proposal of the same year, focused on healthy life expectancy, the possibility of biomedical technology to further it and the economic scenarios associated with these possibilities. On activity, the Committee chose to complement ongoing work on working life with a specific examination of the role of unpaid activities in older people's wellbeing and the wider society (Committee on an Aging Society, 1985). On the built environment, the Committee explored the technological and policy implication of designing housing, transportation and tools to support and enhance older people's daily activities (Committee on an Aging Society, 1988; see also Chapter 8).

Health, activity and technology became, through these and other initiatives, the main transversal pillars of contemporary research agendas for the ageing society. It is thus not difficult to identify an affiliation between these early attempts to reduce uncertainty by settling on a set of problems to be investigated, and later, more global pronouncements on the research infrastructure of the 'ageing society'. For example, in the 2007 Research Agenda on Ageing for the Twenty-First Century, already mentioned in the introduction, the UN Programme on Ageing and the IAGG proposed that priorities in research could be decomposed into three key areas. The first aimed to understand the 'productive contribution of older persons [...] to the social, cultural, spiritual and economic "capital" of all nations' in relation to changing social security provisions and family structures (UNPoA/IAGG, 2007: 3). The second applied to issues to do with 'healthy ageing' and how it related to biomedical intervention and public health measures. The third concerned 'ensuring enabling and supportive environments' – physical and social – for active and healthy ageing.

It is possible to think of these three areas as being related through a loosely coupled triangle. To think of the 'health' axis, we might want to emphasise how practices of health production and measurement become linked to technoscientific promises of recomposition and regeneration of the ageing body (Moreira and Palladino, 2005, 2008), and how these might, in some instances, be configured as forms of biocapital (Rose, 2009; Cooper, 2011). Taking work and activity as a point of entry would entail exploring how re-evaluations of the value of labour and its relation with ageing are entangled with health production, on the one hand, and 'supportive environments' on the other (see Chapter 7). Focusing on technology requires, in the first instance, understanding how through epidemiology and health research relations previously labelled as 'social' – families, friendships, meals, etc. – become qualified as salutary or prejudicial, leading to attempts to instrumentalise and manipulate them. This extension of the range of techniques or interventions blurs, in the case of ageing, the distinctions between biomedicine and public health that have prevailed in other domains (Butler *et al.*, 2008; see also Chapter 9). Technoscientific promises in the domain of ageing offer to modify health and work through a set of converging tools and forms of knowledge that range from the molecular to the sociological.

From this perspective, the 'ageing society' is an assemblage that constrains uncertainties brought to bear with the crisis of actuarial projections. It can be thought of as a way of framing the organisation of knowledge production, technological development and policy formulation that is required by the indeterminate future the 'ageing society' formulates. A measure of its pervasiveness is the fact that the debate between pessimistic and optimistic takes on demographic ageing hinges on differences in the extent and character of the relationship between health, work and technology, and their consequences for economic policy. Further, an indication of its stability as a mode of organising the nexus of ageing and technoscience is that it has remained more or less unchanged for the last three decades. To understand what enabled and sustained the 'ageing society' gathering into a thing, it is necessary to look deeper into the various components within each area of the triangular relationship. This will be the main aim of the following chapters.

Chapter 4

The 'ageing society' and its others

Introduction

In the previous chapter, I proposed that the 'ageing society' should be thought of as an epistemic assemblage. In this, I foregrounded the knowledge making procedures and institutions, the techniques and technologies that shaped how we come to see our societies through a demographic, population prism. I argued that there were two constituents in this assemblage: one related to how the actuarial sciences came to mediate the relationship between the mechanism of life and the economy in 'insurance societies', the second related to how demographic and actuarial forecasts were called into question and prompted a collective search for a new epistemic architecture that was not only concerned with fertility and mortality but also with issues of health production and the 'environmental' manipulation of functionality.

But why was further knowledge on health and productivity seen as the solution to the 'ageing society'? This might seem like an odd question given the societal normative consensus on the value of health and wellbeing for individuals and populations. Drawing on Foucault's writing on biopolitics, explored in the last chapter, scholars have proposed that advanced liberal democracies have experienced shifts in the way the 'mechanisms of life' are known and managed. Rose (2009), for example, has suggested that, in contemporary societies, biopolitics has increasingly focused on the deployment of 'human vitality and morbidity' in detriment of problematisations of mortality, in what he called a 'politics of life itself' (also Rabinow and Rose, 2006; Wahlberg and Rose, 2015). Palladino and I (2008) have argued however that instead of positing a substitution of life for death, we should conceive of those terms entering a new relationship.

This complementary viewpoint is of crucial importance when investigating the processes through which solutions to the 'ageing society' have been so strongly linked to health, vitality, productivity, etc. Indeed, tracing these processes, as I will show in this chapter, requires a re-examination of how we conceptualise the 'population problem' in relation to the 'ageing society'. In demography, population dynamics is usually seen as resulting from the interaction of births, deaths and migration. Yet in scholarship on the biopolitical, concerns with migration and border control are usually detached from those focused on the 'politics of life itself'. In Foucault's original lectures on biopolitics, however, there is a strong indication that death, life and migration were equivalent constituents of the problem of circulation. Such an approach is not only more adequate to analyse the knowledge practices of population scientists but is also crucial to explore the 'nexus of ageing and migration as a governance hotspot for the wealthy capitalist world' (Neilson, 2003: 176).

This chapter aims to rethink how the relationship between birth, death and migration has shaped the epistemic assemblage of the 'ageing society'. To do so, I explore how the stabilisation of the link between the mechanism of life and the economy relied upon what I label a eugenicist agencement focused on the 'quality of population'. I argue that this epistemic and policy concern with 'race' and migratory flows had a significant role in shaping social security systems in liberal welfare states. I trace the transformation of population science in the mid years of the twentieth century, by focusing on how it experienced a shift away from lowering birth rates, which it considered key to economic development, towards a focus on life expectancy and epidemiological aspects of illness onset and management across the life course. I suggest that this focus on health and productivity was achieved through a reinforcement of the problematic role of migration in the cultural economy of population management, where its quantitative redress of population ageing became increasingly trumped by the aim to bolster the use of existing 'human capital', through the extension of health and active working life. My argument is that race and migration can be seen as what Law (2004) would label as the 'absent other' in the constitution of the 'ageing society', that is to say, a set of entities and relations the marginalisation of which is essential for the epistemic and political formatting of an issue (see also Moreira, 2012b).

Retracing the population problem

In tracing the configuration of the 'ageing society' as a population problem, I have argued, in the previous chapter, that it was important to complement Foucault's analysis of 'the set of mechanisms through which the basic biological features of the human species became the object of a political strategy' (Foucault, 2009: 1) with a focus on the work of Thomas Malthus. By positing scarcity as the key nexus through which the mechanism of life relates to the economy, Malthus' model helps us understand how the problem of population dynamics – growth and decline – became central to a biopolitics of security, as an essential component of states' planning for war, taxation and other administrative purposes. Thus, in what is considered one of the first models of population growth, Verhulst, working with Quetelet for the Belgian state, had proposed that population growth was governed by what he called a 'logistic' function, whereby fertility would be limited by the effect of a 'carrying capacity' of a particular territory (Kingsland, 1982).

In proposing this model – which later would instigate, with much controversy, Pearl to characterise it as a 'law of population growth' (Ramsden, 2002) – Verhulst had excluded migration from the equation. This was most likely related to what Caestecker (2000) characterised as a 'brutal' policy of 'alien' control in the Belgian territory in those decades, in that it attempted to model the dynamics of the 'native' population in isolation. The significance of the police apparatus to control 'migration flows' is revealed in the contrast between Verhulst's approach to population dynamics and that deployed in the US where, during the nineteenth century, various waves of immigration came to shape discussions on the relationship between population and the economy. In this, the key document appears to be the publication of the first *Statistical Atlas of the United States* in 1870.

This document is relevant for our purposes because it contains the first recorded use of the population pyramid, the bilateral histogram, which is pervasive in assessments of population ageing. The innovative character of the *Atlas* is usually seen as the result of the military techniques and approach implemented by its director, Francis Amasa Walker. Born in 1840 into the New England social and cultural elite, Amasa Walker taught political economy before becoming an officer during the Civil War. This combination of expertises led him to the position of Chief of the Bureau of Statistics in the Treasury in 1869. Alternating academic and statistician roles, he came later to serve as the American Economic Association's first president, signalling his position within academic and policy circles in the late nineteenth century.

Detailing Walker's work in the organisation of the US Census, Hannah (2000) argues that this relied on establishing an infrastructure of information access and collection. This, in turn, was underpinned by two sets of practices: first, the drawing of grids of reference, supported by military enforcement of access to land, and second, the conception of techniques of 'enumeration' and control of data reliability. These techniques were introduced as a response to what Amasa Walker saw as challenges to the statistical instrument brought by the processes of industrialisation, immigration and urbanisation. By introducing mobility and nimbleness, modernising flows required an infrastructure supported by the operation of 'centres of calculation', with efficient and technically qualified record creation and keeping, and overall rational processes of information management. Chief among these flows was, in the 1870s, the incoming of Eastern and Southern Europeans to the US territory, making, for statistical calculation, race and nativity 'the principal elements of the population' (Walker, 1874: 1).

It is this nativity-centred enactment of population, rather than a concern with 'ageing', that enables us to understand Walker's use of the population pyramid in the *Statistical Atlas* (Figure 4.1). As a means of comparing 'native' (i.e. white north European) with 'foreign' – Southern and Eastern European (Walker, 1874: 39) – populations, the population pyramid was an ideal 'immutable immobile' (Latour, 1990), enabling the observer to identify 'at a glance' drivers of population growth. In the event, what these comparisons made mostly visible was the

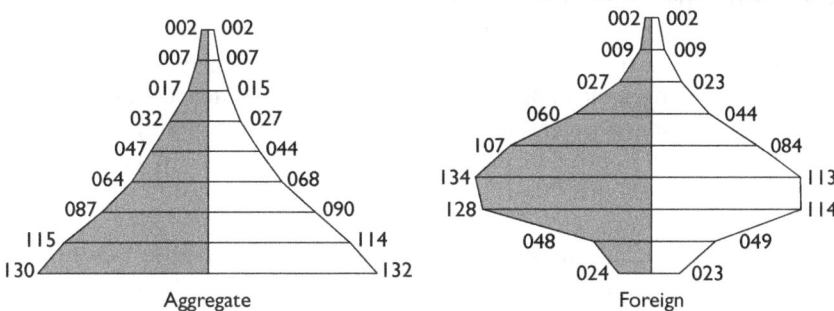

Figure 4.1 Walker, 1874: 39.

effect of recent immigration on what Walker conceptualised as 'the political and moral constitution' of the nation (Walker, 1874: 5). They would also prove crucial in shaping Walker's view of the relationship between population and economic processes.

The population pyramids revealed that a steady growth in the US population in the mid years of the nineteenth century had been achieved despite declines in the fertility of those whom Walker classified as 'native'. The oddity was explained, the graphs made clear, by the immigration influxes that had occurred between 1830 and 1850. Contrasting this more recent period with 'the time when our population was purest, when immigration was [...] hardly appreciable [and] the American people [showed] the capability of doubling their numbers in 22 years' (Walker, 1873: 44), Walker slowly came to the idea that the decline of the 'natives' could be *explained* by the arrival of the 'foreigners'. This was to be crystallised years later in Walker's 'race suicide theory'. Race suicide was the idea that immigration constituted, as he put it, 'a shock to the principle of population among the native element' (Walker, 1891: 632; see also Walker, 1899), in that it had activated one of the Malthusian 'positive checks' whereby 'natives' would restrain from reproducing. This process was one of race self-preservation mediated by economic mechanisms. As Walker himself explained,

> The American shrank from the industrial competition thrust upon him [...]. He was unwilling himself to engage with the lowest kind of day labour with these new elements of the population; he was even more unwilling to bring sons and daughters into the world to enter that competition.
>
> (Walker, 1896: 828)

Walker identified the working of competitive forces in the labour market, underpinned by the mobility of immigrants, as leading to an undervaluation of 'native' labour. Under such conditions, the 'native element' would withdraw its and its offspring's labour from the market with consequences for the 'moral and political constitution of the nation'. This sense that the forces of competition worked against the interest of the American nation helps explain the increased involvement of Walker in policy making towards the end of his life. The 'mental and moral disease' brought about by unrestricted immigration could only be addressed by applying a scientific approach to population policy, a policy 'enforced by the whole power of the state for the good of all' (Walker, 1899: 469). It was in eugenics that Amasa Walker found the basis for this scientific approach to policy.

Generally speaking, eugenics was an intellectual and scientific movement established in the later decades of the nineteenth century, which aimed to improve the quality of populations by the control of human breeding. This approach was based on the idea that differences in abilities, character or temperament are caused by variances in hereditary traits, normally associated with human groupings such as 'races' or 'classes'. Francis Galton, the founder of modern eugenics, proposed that this was to be pursued by a two-pronged approach whereby measures would be used to, first, 'check the birth rate of the unfit instead of allowing them to come into being', followed 'by furthering the productivity of the fit by early marriages and the healthful rearing of children' (Galton, 1904: 23). While nowadays eugenics is mostly associated with the policies of negative 'racial hygiene' pursued in Nazi Germany, research in the history of economics has begun to detail how eugenic thinking came to shape the relationship between population science and economics at the turn of the twentieth century.

Leonard (2005) has suggested that, in the US, such development was underpinned by emerging concerns about immigration. In this, economists like Amasa Walker were able to convert social anxieties about incoming Southern and Eastern Europeans into an expert discourse about 'race suicide', a concept promulgated by central figures in economic thinking such as Frank Fetter or Irving Fisher. The idea of 'race suicide' represented a departure from the 'laissez-faire' economics favoured by liberal economists, and argued for interventionist population policies led by the state. Like Amasa Walker, other US economists saw this as resulting from an unintended consequence of industrial capitalism in that it had created the 'artificial' conditions for the 'unchecked' reproduction of 'defectives, delinquents and dependents' (Fisher in Leonard, 2005: 209). Some, like the sociologist Edward Ross, went so far as to suggest that the immigrant inferior races were more suited to the 'insalubrious' condition of urban, industrial living than Anglo-Saxon stock – a niche advantage that it was necessary to correct through rational policy.

The need for addressing the weaknesses of liberal political economy focused also on the idea of the *homo economicus*, i.e. the idealised average human being capable of taking rational decisions. This was particularly significant in British economic thinking in trying to make sense of the problem of the 'industrial classes'. In this respect, Levy and Peart (2004) document how eugenic thinking penetrated British economics at the turn of the twentieth century by articulating a connection between the biological and economic qualities of the poor. Again, the evidence mobilised for this connection was the way in which high, unchecked fertility rates denoted an incapacity for restraint and foresight that was characteristic of the 'economic man'. This created problems for economic theorising in that the assumption that the economy was made up of 'rational men' was becoming undermined by demographic reality. As the leading British economist Alfred Marshall put it in his *Principles of Political Economy*, there were 'increasing reasons for fearing that the progress of medical science and sanitation is saving from death a continually increasing number of the children of those who are feeble physically and mentally' (Marshall, 1890: 201). Again, here also it was immigration that underpinned such anxieties, identifying the arrival of Irish immigrants and Eastern European Jews as a threat to the genetic quality of the British population.

Nowhere is the centrality of eugenics for population and economics thinking more clear than in the work of Carr-Saunders (Osborne and Rose, 2008). A disciple of Karl Pearson at the Galton Eugenics Laboratory, Carr-Saunders came to work closely with Julian Huxley, one of the central proponents of the modern 'synthesis' of Darwinism and Mendelianism, in developing his approach to 'the population problem'. He proposed the view that while most public focus about population after World War I had been related to a fear of demographic and political decline, there were deeper issues concerning the relationship between the 'quantity' and the 'quality' of the population. In this, Carr-Saunders questioned whether a strict Malthusian model by itself could guide thinking about the 'population optimum':

> There does not seem to be any reason for supposing that there is any limit to the increase of skill in food production, and that therefore there is any limit to the desirable number so long as the criterion remains economic. This suggests that at some point mankind will have to introduce another method of estimating what density is desirable, as it is clear that the economic advantages of increase somewhere come into conflict with other ideals as to desirable social conditions.
>
> (Carr-Saunders, 1922: 309)

With this, Carr-Saunders was directly challenging the Malthusian argument that population growth would be checked by the rate of increase in food production. In the absence of such 'natural' processes of determining the population optimum, other criteria would have to be introduced. The search for other criteria was underpinned by Carr-Saunders' eugenicist belief that while the English upper and professional classes had grasped the advantages of the effect of low fertility and self-restraint on their living conditions, at the 'bottom of the social scale' this was not the case. The population problem arose from these high reproductive rates among 'primitive people' with low mental and physical qualities.

This led to fundamental changes in the composition of the population, in the making of what Quetelet once named the 'average man', which endangered the standard of living of the nation. There was thus a negative, reinforcing loop between the creation of wealth and the 'quality of the population' which required changes in policy in education, birth control, etc., so that 'innate mental differences manifest themselves as between man and man' (Carr-Saunders, 1922: 428). 'Designs for living' should be directed at maximising the germinal potential, working with natural processes to incentivise the identification and rewarding of natural talent. Indeed, Carr-Saunders' argument fully justified, on the basis of the genetics of population, the liberal welfare state, and was most certainly the basis for his appointment to the Charles Booth Chair in Social Science at the University of Liverpool in 1923.

Carr-Saunders' public recognition is part and parcel of the establishment of social science in the inter-war years. As Leonard (2005) demonstrates, sociologists had been at the forefront of the eugenicist critique of classical political economy from early on, which was advocated not only by Fabian Society founders Sidney and Beatrice Webb but also by more established academic figures such as Chicago sociologist Charles Henderson, or the first professor of sociology to be appointed in the United Kingdom, L. T. Hobhouse – to whom Carr-Saunders dedicates his 1922 opus. It is also evident in British economist and political reformer William Beveridge's attempt to create a discipline of Social Biology while directing the LSE in the 1920s (Renwick, 2012). This tradition of research was consistently motivated by an interest in the relation between 'natural ability' and the social system of classification, which had been sparked by Galton's study of the relationship between eugenic value and 'social class', marking a 'depart from the old sociology' and the creation of a 'new calculus' (Pearson, 1909: 20; MacKenzie, 1976; Thévenot, 1990). It was a search for a sociotechnical agencement that would best articulate social institutions – Frank's 'designs for living' – and nature.

Population and the eugenic 'agencement' of the liberal welfare state

One of the significant consequences of the penetration of eugenics in economics, social science and population science was, as we saw in the last section, the establishment of a call for a more scientific – natural-fact based – approach to public policy. This entailed an exploration of new articulations between nature and society, and a questioning of 'old' laws, predictions and forecasts. It was not only that the Malthusian 'principle of population' failed to accurately determine the size of the population; it was also that it was not enough to ensure the 'quality of the population'. However, if laissez-faire, competition-based policies were not able to deliver the 'common good', what other basis could be used for the design and implementation of policies? How could administrations distinguish between worthy and unworthy causes and people? Who was to say and who was to benefit?

In the US, answers to these questions coalesced in the political movement that is normally labelled the Progressive Era, a period characterised by increased involvement of the state in social affairs, which saw the creation of the liberal welfare state, with personal income tax, the Federal Reserve and strengthened regulation of immigration. As was suggested above, and in the previous chapter, the technical formulation of such policies was heavily reliant on the expertise of social and population scientists. In this, as we saw, eugenics – and the question of the 'quality of the population' – played a fundamental role, helping to anchor policy making in education, health or immigration. A similar although less distinct period of knowledge-driven political reform took place in Britain. This is often associated with the Liberal reforms of 1906–1914 led by Lloyd George, already touched upon in the last chapter. As in the US, eugenics became a key mediator in the relationship between experts and policy makers, providing a basis for evaluation of Liberal policies on income tax, family policy, education and poverty. But nowhere is this eugenic anchoring of policy more rooted than in the framing of the Old Age Pension reform of 1908.

This claim might seem unusual given that both negative and positive eugenics aim to shape the genetic pool of a population, a focus which would ostensibly exclude those outside reproductive ages. However, in analysing political debates around the introduction of old age pensions it is possible to identify a concern with the 'quality of population' in allusions to 'deservedness' and unease about the effect of pensions on the social control of productive behaviour (Thane, 2000: 195–255). Prompted by interest about the fate of the 'aged deserving poor', following the publication of Booth's 1890 survey, the political debate shared with Booth's study its normative valuation of people, according to their willingness to work, aspiration and capacity for self-care and planning. Drafted in close consultation with the Webbs, the 1908 old age pension scheme, as was referred to in the last chapter, explicitly aimed to incentivise industrious engagement with work and parsimony with money by excluding those who had shown 'an habitual failure to work', alongside those convicted for crime or drunkenness, and 'aliens' (Thane, 2000: 218–220). The aggregation of these categories is redolent of the eugenic imaginary by the drawing of a semblance between the 'bottom of the social scale', moral weakness and inferior races, thus preventing the 'country gradually falling to the Irish and the Jews' (Sidney Webb in Semmel, 1958: 116).

While in the US and the UK eugenics never became part of the normalised policy instrumentarium, in the same way that countries like Sweden, Denmark or Norway based state intervention explicitly on ideals of national productivity (Broberg and Roll-Hansen, 1996; Spektorowski and Ireni-Saban, 2014), concerns with the 'quality of population' became embedded in the development of the welfare state in the first four decades of the twentieth century. Indeed, such concerns pervaded the formulation of policies seemingly not related to eugenic valuation, such as old age pensions. The establishment of this direct link between population and economic value is usefully embodied in the figure of William

Beveridge. An economist by training and an active political figure in the Liberal Party, Beveridge's distancing from laissez-faire politics and openness to state intervention appears to have developed through meeting the Webbs in the first decade of the 1900s (Harris, 1997[1977]: 87), and his visit to study the German old age pension scheme (see Chapter 3). Such experiences seemed to have cemented his belief that 'social Institutions should be designed to work with nature rather than against her' (Beveridge, 1943: 161), a belief that presumably motivated his interest in Social Biology (see above) but most importantly helped formulate what is considered the key blueprint for the modern, post-war welfare state. Thus, when contextualising what became known as the Beveridge Report to the Eugenics Society, Beveridge argued that,

> [W]ith its present rate of reproduction, the British race cannot continue and that 'means of reversing the recent course of the birth-rate must be found' [and thus the Report proposed that] children's allowances can help restore the birth rate by making it possible for parents who desire more children to bring them in the world without damaging the chances of those already born.
>
> (Beveridge, 1943: 151)

For Beveridge, the issue of the decline of the population – or demographic ageing – seemed as important as the 'quality of the population', particularly as the former was linked to the availability of recruits during the war, and concerns about economic power. However, for him, one would not have trumped the other, and policies could be designed that aimed to increase the birth rate by facilitating the realisation of the desire to have children among 'the most capable', who would otherwise have 'fewer children'. Emphasising that his was a 'Plan on British lines', Beveridge argued that children's allowances would not 'favour breeding from those who are less successful' but rather incentivise the reproduction of those who otherwise refrained from additional reproduction to maximise their, and their children's, economic chances.

This example makes clear why historian of eugenics Diane Paul (1995, 2001) argued that the full political potential of the intellectual and scientific concern with the 'quality of population' was only realised with the establishment of the modern welfare state. It was only within the context of the extension of the regulatory powers of the state over the labour market, and the comprehensive involvement of central and local administrations on education, health, old age pensions and other welfare services, that it was possible to bring to bear a combination of technical expertise and decision-making capacity on managing the quantity and quality of the population. Again, this combination was operational both in more interventionist, social-democratic welfare states such as Sweden or Denmark, and in those more concerned with balancing individual freedom and the collective good. In both contexts, the knowledge created by economists, population scientists and, increasingly, sociologists became key in shaping and legitimising new institutions and policies. As Myrdal would put it: 'We have today in social science a greater faith in the improvability of man and society than we have ever had since the Enlightenment' (Myrdal, 1944: 1024).

It is generally argued that, with the uncovering of the human impact of the Nazi policies of 'racial hygiene', eugenics fell out of favour in mainstream intellectual and political circles. Eugenics became mostly associated with its negative policies, while positive, incentivising measures such as child benefit or old age pensions became increasingly associated with family policy, child health and ideals of 'social justice', fairness and opportunity. However, it is possible to argue not only that concerns with the quality of population scaffolded much of the formulation of modern welfare policies but also that there is a multi-layered continuity between the eugenic concern with the 'quality of the population' and the welfare state's aim of building and sustaining a national, bounded *community*. In this national community, individuals identify with each other and share common goals across social class, gender or generation, a solidarity that, as Beveridge argued in relation to the German old age pension, would be mediated by the state (see Chapter 3). Indeed, early welfare state theorists knew reforms such as old age pensions would only be sustainable through the construction of what Anderson (1983) labelled 'imagined communities' (see Wolfe and Lausen, 1997).

The 'ageing society' and the reconfiguration of productivity

The sustained expansion of old age pensions and services specifically aimed at older people in the years after World War II marks the transition to a modern welfare state, where concerns with social justice and national solidarity are emphasised in detriment of strict productivity concerns. This relied, as I argued above, on creating and maintaining economic and cultural bonds between citizens (Marshall, 1953). There were, however, key changes in the understanding and management of the populations in the post-war period that would challenge the circumscribed, national nature of such bonds. In this, a crucial event had to do with what the historical sociologist Charles Tilly (1978) once described as 'one of the greatest whirlwinds to sweep the earth', referring to the scale of displaced populations during and after World War II. Because these movements were, to a large extent, negotiated, administered and coordinated by central, shared authorities, they can be seen to have led directly to the creation of the United Nations Populations Division (UNPD) in 1946. This institution extended the techniques and practices of the census geographically, adding to the infrastructural underpinning of demography as a scientific, international discipline, by enabling a monitoring and comparability of population dynamics in different parts of the world. It could be seen as the birth of an international biopolitics (Wahlberg and Rose, 2015).

Second, but intimately related to the previous point, was a complex shift in populations science, in how the birth rate was seen to relate to socio-economic conditions. This is epitomised by the consolidation of what many consider to be the central scientific paradigm in population science: demographic transition theory (e.g. MacInnes and Pérez Díaz, 2009). As Ramsden (2002) has shown, demographic transition theory represented a direct challenge to models of population dynamics that relied on actuarial functions of mortality such as those of Gompertz (Chapter 3), and in particular to Pearl's use of Gompertz to propose the view that populations were governed by 'biological laws'. Drawing on earlier work by Thompson (1929) on different types of population dynamics and their relationship with territories, transition theory stipulated that socio-economic change – industrialisation – prompted population growth by reducing mortality while high levels of fertility were maintained. These levels of fertility were predicted to decrease as the culture and habits of restraint and planning – of Malthusian thrift – spread throughout the population, leading to decreases in the number of births, and to a state of 'low level equilibrium', or 'incipient decline' in the population.

However, as Szreter (1993) suggests, while the original models proposed by Notestein and Davis in the 1940s proposed that fertility should be seen as the ultimate dependent variable, by the 1950s fertility had become an independent, plastic variable, used to activate social and economic modernisation in non-industrialised countries. This transformation had been effectively mediated by the increased penetration of the UN and other international organisations into the management of populations, referred above. In this context of international biopolitics, technological control of fertility in local populations was intimately linked to the 'cold war' contest for power and influence over what French demographer Alfred Sauvy (1961) labelled the 'Third world'. Transition theory became thus both a political technology and a technology of politics, as it enrolled the Ford and Rockefeller Foundations' interest in funding family planning programmes through the UN (Connelly, 2009), and in making such programmes the unique pathway to capitalist economic development at least until the 1980s.

These geopolitical processes were concurrently related to the consolidation of a neo-Malthusian approach to population growth that saw it in contradiction with the ecological dynamics of the planet. Mostly associated with Ehrlich's (1968) *The Population Bomb*, it drew explicitly on Verhulst's notion of 'carrying capacity' (see above) to argue for imposing limits to growth (Dean, 2015). Moving this notion from a national to a planetary level, it further reinforced the need for population control, managed through a focus on 'shared resources' and institutions of international governance. These assumptions became crystallised in the notion of 'sustainable development' in the current discourse of mainstream environmentalism. From this perspective, fertility became not only an economic technology, but was also understood as a socionatural implement, whereby the managing of resources for 'future generations' progressively impends on the present situation.

This shift in the epistemic position of fertility in population dynamics is significant because, in industrialised countries, it became instead associated with a decrease in focus on maintaining replacement levels in the population. As was discussed above, while previous fertility incentivising policies – child benefit – had sparked concerns about the 'quality of population', they also represented the only sure path to economic growth (e.g. Keynes, 1937). The problem was that, in the decades after the war, raising levels of fertility would, in the short or medium term, lead to an increased burden on the productivity of the active population, in a context of expansion of child and education services. Incentivising fertility carried its own risks within the 'insurance society'. Other solutions were necessary.

As is well known, in European countries which were pursuing active policies of economic restructuration such as Germany, France or Britain, solutions to the problems of labour shortage came in the form of 'managed immigration' programmes. These were centrally administered schemes of recruitment, selection and placement that supported the flow of 'colonial populations' to the metropolis, and incentivised foreign workers to find employment in the host countries (Goldin *et al.*, 2012: 85–90). To be fair, it was known that immigrants would also incur consumption costs in welfare and services in the short term (see below). These were, however, thought to be comparatively lower than those sustained by child rearing and education, and were offset if the productivity of incoming workers was insured by training and rational assignment of employment. Immigration presented a quick answer to labour shortages, but was epistemically and politically problematic. This troubling character of migratory flows only became fully evident as demographic ageing was delineated as a bounded research topic in the post-war context (Rosset, 1964).

The process of formation of this subfield of population research was decisively shaped by the work of Jean Daric, a young demographer working with Sauvy in the then recently created French Institut National d'Études Démographiques (INED). In a series of papers published after 1946, Daric aimed to rearticulate Carr-Saunders' – and other 'social darwinists', as Daric saw them – view that eugenic values should guide reasoning on the economics of population. Rather than pursuing a population optimum derived from a trade-off between quantity and quality, Daric posited that this should be underpinned by the question of maintaining the population's 'standard of living' – *niveau de vie* (Daric, 1946: 71; Daric, 1952). Such an economic criterion was not simply a version of quantity, as previous models had suggested, but a reconfigured version of 'quality', and a new practice of valuation.

Daric suggested that the main challenges to French citizens' standard of living were demographic through and through. They were related to the long-standing process of population decline, and the probable reduction in the proportion of the active population in the future (Daric and Girard, 1948). He proposed three ways of maintaining this general, average standard of living. One would entail increases in the efficiency of labour by technological innovation. The second involved a sustained growth in the number of active persons through immigration. The third was the prolongation of active life, through the postponement of retirement. Daric argued that, while important, the first solution could only go so far in solving the problem of labour shortages in the future. The second possibility was more problematic because, Daric suggested, while immigration had had a crucial historical role in slowing the process of population ageing in France and other countries, it also presented political, economic and social risks 'to an ageing and impoverished country' (Daric, 1946: 73). These risks were related to what Sauvy (1950) – Daric's supervisor at the INED – called the 'national interest' and the 'cultural unity' of the nation. Although redolent of Amasa Walker's construal of the threat of immigration to 'the political and moral constitution' of the nation, in the post-war period such concerns, as I argued above, were cast in a cultural language that appealed to the solidarity mechanisms of nations identified by Durkheim, and sought by policy makers like Beveridge (see Chapter 3).

Because of the uncertainty of technological innovation and the disquieting effect of immigration on 'imagined communities', Daric proposed that the only realistic solution to the 'ageing society' would be exploring ways in which we can make 'a *better use of our own human capital*' (Daric, 1946: 73, my emphasis). This statement is of crucial importance in tracing the assemblage of the ageing society. It marks a shift in what I called the eugenic agencement of the 'ageing society', from a fertility-led to a mortality-focused management of population, in which migration ceased to be a problem for the genetic quality of the population and became a cultural and social challenge to the apparatus of the 'insurance society'. Importantly, in this transition, the question of productivity remained central to population reasoning and knowledge making of demographic ageing. To scaffold his argument, Daric uses a term – human capital – that has become associated, through Foucault (2009), with the emergence of neoliberal economic models and forms of governance. However, the idea of human capital – or the productive value of labour – had been a consistent focus both of economic reasoning and in the insurance industry, with Marshall, already in the early 20th century, for example, suggesting that the value added by an immigrant to the economy should be calculated as a 'discounted' price of the relationship between production, consumption and length of life (Marshall in Hofflander, 1966).

On the other hand, in the context of post-war societies, the transition to retirement could be thought of as an enhancement of demand-led economic growth, a process usually associated with Keynesian economics. However, in welfare policy and economic calculations, older people weren't yet fully accepted as part of consumption-led incentives to productivity, being cast as 'dependent' on productive labour, or living below the 'standard of living' (Townsend, 1957). This meant that reassessments of the economic value of older people had to rely only on gains in productivity, which, in the absence of definite innovation-based improvements, could only come from an extension of working life (Daric, 1952). Such a complex configuration of elements worked together to make the question of the relationships between age and productivity crucial to technological democracies, as discussed in the last chapter (see also Chapter 7). Indeed, Rosset, in an exhaustive critical review of research on 'ageing populations' in the two decades after World War II, hypothesised that a re-evaluation of the value and usefulness of older workers might be nothing more than 'a natural consequence of the demographic changes that have been taking place' (Rosset, 1964: 302).

This hypothesis was, in some ways, a logical corollary of locking-in age grading in the economic assumptions of old age pensions, as discussed in the last chapter. Indeed, the members of the US Committee on Population Problems – established to understand the consequence of the social security reforms of 1935 – had already suggested that fertility and immigration were unlikely to be useful tools in economic policy in the future (Hoff, 2012: 96–104). From the Committee's perspective, existing forecasts of an increasingly stabilised, *ageing* population – heavily reliant on Warren Thompson's (1929) questioning of the Malthusian fixed relationship between fertility and mortality – precluded the customary quantitative solutions of economic growth. In populations with concurrent declining birth and death rates, gains in productivity would have to be made at the later part of life. Such population scenarios would require, in the view of their expert members – including Thompson himself – 'an adaptation of work to older workers' (Committee on Population Problems, 1938: 65) in order to ensure the maintenance or enhancement of what Daric later called the 'standard of living' (see also Chapter 7).

Equally, in proposing his old age security reform in Britain, Beveridge had been keen to remark that the 'object of the scheme isn't to force early retirement but to leave men (*sic*) to retire when they want and to encourage them to go on working while they can' (Beveridge, 1942: 55). In making this statement, Beveridge was articulating both his political commitment to supporting thriftiness, and an awareness of the financial implications of a pension scheme implemented within a declining, ageing population. The actuaries' assessment of the pension scheme implemented after World War II, drawing on the Beveridge Report, came to similar conclusions, arguing that,

> it is clear that the only way in which the growth in pensions can be provided, without reducing either the standard of living of the rest of the population, or the amount devoted to new investment, is to increase either the productivity per head of the working population or to increase their numbers by postponing the normal ages of retirement.
>
> (Bacon *et al.*: 1954: 155)

In effect, what pension schemes implemented in Britain, France, Germany or the US brought to bear was that productivity gains were the key path for economic growth within stable or declining populations. In this, the consensus appeared to be that if standards of living were to be maintained, the conditions upon which these schemes were constructed, namely a minimum retirement age, were unsustainable. It was necessary to devise ways to make, as Jean Daric put it, 'a better use of our own human capital', either through technology or longer working lives. The 'ageing society' called for human ingenuity and innovation;

writing into its constitution the need for scientific and technological research on ageing. It is from this perspective that we can understand why health, and the technoscientific manipulations of functionality, became central to the assembling of the ageing society. It was the result of a series of epistemic and policy lock-in processes, whereby raising the efficiency and productivity of existing human capital became the only possible, imaginable pathway for managing the economics of ageing populations.

This was not without its problems, as one of the main challenges in raising productivity in an ageing population was the established consensus that ageing was associated with physical and mental decline. As Katz (1996) has shown, this conviction had been fundamental to the disciplinary formation of knowledge-power on ageing between the nineteenth and twentieth centuries. Economic concerns with ageing populations, already present in Amasa Walker's comparison between 'native' and 'foreign' age structures, were also based on the same belief that it entailed a decrease in aggregate productivity. Daric himself, while calling for an extension of working life, was convinced that 'efficiency of production [was] inversely connected with age' (Daric in Rosset, 1964: 302). To be exact, however, demographic transition theory, for all its focus on the socio-economic aspects of population change, said nothing of the consequences of what it saw as a stable population on productivity, and the value of older people's labour (Cowgill, 1974). To be able to address this problem, population science had to engage with questions of morbidity, and its relation to mortality.

Vitality and the epidemiological transitions of the global society

In the previous section, I suggested that the ability to gather data and compare across populations had been fundamental for shifting the epistemic position of fertility in population models related to demographic transition theory. This form of international biopolitics enabled and was supported by interventions on family planning which were seen as routes to producing wealth creating families through Malthusian thrift. As part of this apparatus, international agencies such as the UN developed expertise and capacity on managing family reproductive life. Such practices, however, presented a different angle on the mechanisms of life, because they were removed from the measurement and the ultimate goal of economic growth, and dealt mainly with prevention of child and maternal mortality.

It is thus not surprising that it was within this context that a response to demographic transition theory first emerged: the theory of epidemiological transition (Omran, 1971). The theory echoed the developmental narrative of transition theory by also postulating three stages in population dynamics. The 'age of pestilence and famine' was characterised by high mortality and no population growth; in the 'age of receding pandemics', decreases in mortality led to exponential population growth; and finally, the 'age of degenerative and man-made diseases' was defined by an equilibrium in the level and age-incidence of mortality. By focusing on mortality, Omran's theory represented an attempt to claim space for public health academics to speak on population questions, which were then dominated by demographers, sociologists and economists, particularly in settings related to international biopolitics. To do this, Omran drew on his training in health services organisation and delivery to argue that public health infrastructure was integral to the shifts in mortality as conceived by demographic transition theory (Weisz and Olszynko-Gryn, 2010). Because his focus was on the role of health care in developing countries, Omran was not particularly concerned with exploring the consequences of such distribution in population pyramids, or with what Wahlberg and Rose (2015) called the characterisation of 'morbid living'. What his model enabled, however, was a focus on mortality in population research, which until then had been dominated by the driving role of fertility.

Omran was able to categorise historical 'eras' in terms of the dominant cause of death in the population because the 'major premise [of his model was] that mortality is a fundamental factor in population dynamics' (Omran, 1971: 511). By focusing on mortality and how types of disease differently shaped the survival curve in a population, the model identified a long-term redistribution of death from the young to the old. This was the key factor in it becoming a much cited and used paper, because it spoke of the relationship between life expectancy and health that had become a burning issue in the management of the economic consequences of increased life expectancy at the turn of the 1980s, as discussed in the last chapter.

It is also from this perspective that we can understand why Jay Olshansky, himself a social demographer, chose to draw on Omran's epidemiological model to address the questions arising from the 1980s 'social security crisis', discussed in the last chapter. His interest was based on Omran's focus on different 'modes of mortality', because it allowed him to hypothesise that a further shift would occur in age-specific mortality and increases in life expectancy at older ages, in what he and Ault (1986) labelled the 'age of delayed degenerative diseases'. This new era, they argued, was unique because it opened new questions about the biological limit of the human life span, and its interaction with health, ageing and disease. As they themselves put it, these questions were of crucial importance because they all related to the 'central issue [...] of *vitality* – or the question of whether declining mortality in advanced ages will result in additional years of health or additional years of senescence' (Olshansky and Ault, 1986: 379; original emphasis).

The problem was that not only this latter question very much undecided, but there was no agreement on how to measure and understand the relationship between mortality and morbidity. Would gains in health be reflected on a 'squaring of the curve' of mortality as Fries' (1980) compression of morbidity hypothesis would imply, or would comprehensive shifts in age-specific mortality mask an extension of periods of disability (Schneider and Brody, 1983)? Such questions reinforced the need for reliable and practical measures of morbidity or disability that were independent of age (see Chapter 5). This need had first become salient with the establishment of health programmes such as the National Health Service in Britain, or later in public debates about the extent of a federally funded programme in the US, eventually settled in the creation of Medicare (Oberlander, 2003). Curiously, however, in these contexts it had been the apparent stability of mortality rates in the post-war years which justified the creation of new metrics focused on the living (Moriyama, 1964, 1968). Supported by developments in computation and sample survey techniques, health measures were thought to enable a valuation of the effects of health care in 'reduced productivity, prolonged disability, and the need for care' (Sullivan, 1966: 1).

When, two decades later, mortality rates were seen to have improved (Chapter 3), mortality and morbidity had been tentatively decoupled. It is important to emphasise the contingent and fragile character of this disjointing because this was exactly the central issue in the controversy about 'vitality'. How to measure this vitality became thus a hotspot in the controversy: should it focus on a medical definition of the normal, or should a more holistic, positive definition of health, as advised by the World Health Organization (WHO), be deployed? Should it be based on objective, physiological measures of function or should it take into consideration the subjective perception of health? Preferences for one or the other were closely associated with what different actors saw as the ultimate goal of health programmes, and their normative grounding. This was further compounded by whether researchers were interested in investigating medical technology, health services or preventive health programmes as 'the relative contributions of each of these circumstances to declining mortality [was] yet to be determined' (Olshansky and Ault, 1986: 360). From this perspective, morbid living (Wahlberg and Rose, 2015) did not replace mortality as the focus of the biopolitical gaze, but opened up death to this very gaze (Moreira and Palladino, 2008).

Wahlberg and Rose are right, however, in linking this interest in the metrics of health to the consolidation of international biopolitics and its transformation into what has become known as 'global health' (see also Adams, 2016). Seen from the focus of this book, what this international biopolitics apparatus enabled was the entanglement of the 'central issue of vitality' with policies and programmes on 'active and healthy ageing' in the 1990s, discussed in Chapter 1. This process can be understood more deeply by focusing on the role of Alexandre Kalache – one of the originators of the concept of healthy ageing – in linking epistemic and policy concerns with the relationship between ageing, mortality and health. Trained as a postgraduate in the academic centre mostly responsible for the proposition of bringing closely together epidemiology and preventive public health, Kalache was a central character in the development of the epidemiology of ageing during the 1980s, and in particular in developing an approach that focused on the ageing of populations in developing countries (Kalache and Gray, 1985).

Of special interest to this research was how countries which in the 1980s had experienced a rapid decline in fertility and mortality rates could manage the process of what Omran had labelled the 'epidemiological transition'. While, as we saw above, the transition had been handled in developed countries through a combination of public health measures, medical innovation and/or health services, developing countries were not seen as capable of extending limited resources – mostly focused on child health – to older people. Drawing on Fries' hypothesis, Kalache proposed that policy makers should draw on public health instruments to lower morbidity and foster the extension of independence in later life, rather than expect rising health care costs as a consequence of ageing societies. Moving from the academic to the policy world in the early 1990s, Kalache's perspective and work were then instrumental in facilitating the globalisation of the 'ageing problem' (Kalache and Kickbush, 1997). This had a crucial effect on how the relationship between births, death and migration was seen to affect 'ageing societies'.

By extending demographic ageing beyond the developed world, 'global ageing' meant that the historical transfers of labour from 'younger' to 'ageing' societies were increasingly seen as politically and economically unsustainable. This combined with one of the key paradoxes of the 'global society' to further exclude – lock out – migration as a factor in the management of demographic ageing. Social scientists usually associate globalisation processes with a contraction of the spatial and temporal dimensions of social life supported by information and communications technologies and transport. And while this transformation has been associated with increased levels of mobility, what is distinctive about the 'global society' is the rising level of state investment in controlling the movement of persons, particular to the Global North (MacInnes and Pérez Díaz, 2009). This is legally epitomised in the European Schengen agreement of 1995 and the creation of what came to be known as 'Fortress Europe'. Such policies and programmes of migration control are meanwhile seen as responsible for the increased wage differentials between the developed and developing countries and reinforcing global inequalities. This means also that increased socio-economic interdependence between North and South has relied on the activities, equipment and technological infrastructure of what some scholars have labelled the 'border machine' (M'charek et al., 2014; see also Walters, 2002; Amoore, 2006; van Houtum, 2010).

The tensions deployed in the management of ageing in the global society came to the fore when, in 2001, demographers at the UN Population Division explored the question of whether immigration could mitigate some of the economic effects of ageing populations. Outlining a number of scenarios regarding the international net migration flows required to achieve particular objectives in terms of age structures and dependency ratios, the report concluded that maintaining 'support ratios at current levels through replacement migration alone seems out of reach, because of the extraordinary number of migrants that would be required' (UNPD, 2001: 4). Instead, the report proposed a five-pronged approach to population ageing that emphasises a societal reassessment of the retirement age and a reorganisation of health and social security schemes. In this regard, the UN Population Division was in keeping with the consensus about the problem of and solutions to population ageing that had been consolidating since after World War II.

Their proposal stated that while migration could reduce the size of dependency ratios, with only a slight growth in the percentage of migrants in relation to levels experienced in the 1990s in Europe or Canada, the number needed to offset long-term demographic ageing, would be 'of substantially larger magnitudes' (UNPD, 2001: 98), which was thought to be politically and culturally challenging for host and sending countries. This analysis of the issue was not dissimilar to one published a few years later, focused solely on the EU, which also underlined 'the *magnitude* of the migration inflows that would be necessary as a supply of labour, in absence of other changes such as increases in the labour force participation' (DG ECFIN, 2012: 54, my emphasis). In this respect, population policy agencies' view of migration flows in the 'global society' was redolent of Daric's focus on the cultural impact of labour replacement programmes, but now it added concerns with the effect of emigration on less developed countries. This meant that migration was not only problematic for the cultural unity of the receiving countries, but also a threat to the political, economic and cultural stability of the 'global society'.

Such locking-out of migration as a solution to the ageing society was further reinforced by responses to the UNPD 2001 report. For example, Coleman, an academic demographer, drawing on an analysis of data from the UK Government Actuary Department, argued that in terms of population dynamics, 'migration has nothing to do with it', adding that,

> In terms of its effects on age-structure, migration is the weak sister of population dynamics and has been relatively ignored as a demographic process. Its unfashionable status arises partly from the multiplicity of numerous definitions compared to the hard biological end points of conventional demography [...]. Technical demographic theory is concerned with stable populations and the rest has been mostly based on closed populations, which facilitate finite solutions.
>
> (Coleman, 2002: 7; see also Coleman, 2000)

In this, Coleman was perhaps alluding to Verhulst's model of population growth, which explicitly – politically – excluded migration. Emphasising only the interaction of fertility and mortality, Coleman argued, was not only advantageous from a mathematical point of view, enabling stable calculations and simulations, but also from the point of view of mensuration, providing 'hard biological end points' for quantification. But this epistemic advantage was compounded by an economic objective, in that the 'reliance upon the apparently easy option of importing labour from overseas' contributed to low productivity and low wages, and prevented structural adjustment regarding postponement of retirement age (Coleman, 2002: 26–28). This last aspect of Coleman's argument is remarkable in that it suggests that immigration undermines the very mechanisms that are consensually agreed to be able to redress the economic consequences of ageing populations, when excluding migration itself. In this, Coleman's reaction to the UNPD report on 'replacement migration' appears thus to revisit the causality arguments put forward by Amasa Walker almost a century earlier in his theory of 'race suicide' (see above), but now adding the role of foreign labour in motivating decisions to abandon the labour market earlier than projected or necessary.

Coleman's intervention, if on the side of those arguing for strong regulation of migration flows, such as Migration Watch – for which Coleman was an adviser – reveals the assumptions he shares with those that he is arguing against. In particular, it identifies the key aspects of the 'nexus between ageing and migration' in the global society (Neilson, 2003, 2009, 2012), in that productivity and wellbeing of ageing populations in the Global North and South are now seen to be directly threatened rather than ameliorated by migration. That is to say, migration has ceased to be simply a possible, if problematic, solution to the ageing society, and has regained the menacing qualities it had in the eugenic agencement by the turn of the twentieth century. Arguing that it no longer impends on the 'quality of population' would be ignoring how the issue of productivity transferred from a germinal concept to 'an understanding of the human body as an assembly of biomolecular components that can be recombined so as to maximise the resultant unit's cultural, social and political productivity' (Moreira and Palladino, 2008: 21; Rose, 2009). In the present situation, strategies to enhance the vitality and productivity of existing 'human capital' appear to be interdependent on the maintenance of biosocial borders, on the 'border machine'.

Conclusion

In this chapter I have analysed how the evolving relationship between birth, death and migration has shaped the assemblage of the 'ageing society'. I have argued that the stabilisation of the link between the mechanism of life and the economy relied upon a eugenicist agencement focused on the 'quality of population', which had a significant impact on the shape of social security systems in liberal welfare states. I then traced the transformation of population science in the mid years of the twentieth century, away from a focus on fertility and towards a focus on the relationship between morbidity and mortality. I suggested that this focus on health and productivity was accomplished through a strengthening of the problematic role of migration in the cultural economy of population management.

In this regard, it can be suggested that the 'othering' of immigration and race in the assembling of the 'ageing society' is a form of epistemic and political 'absent presence' (Law, 2004). This concept describes a relationship whereby entities are excluded from the agencement in the operations of its assembly. They become manifestly absent, explicitly locked out of the assemblage. Their 'out-thereness' is thus a necessary condition for the development and maintenance of the 'ageing society'. Of course, in this case, the locking-out process overlaps with the processes of recording and monitoring the biological characteristics of populations to support the 'border machine'. In social and cultural geography, these processes of management and exclusion of migratory flows have been conceptualised through Agamben's (1998) proposal that the deployment of 'bare life' – outside of the *polis*, epitomised in the concentration or the refugee camp – is integral to the productive powers of biopolitics. While this is an appealing theoretical anchor for the role of migration in the biopolitics of ageing, it does not capture the fluid, difficult to stabilise role that migration had in the articulation of the 'ageing society', serving as a resource for its 'discovery', as a problematic solution, as a moral implement in justifying how institutions should address it, or as a possible threat to those same institutions.

Further, this conceptualisation misses the empirical detail that this shifting position of migration in the assembling of the ageing society was accompanied by an equally important transformation of the role of fertility and mortality in population dynamics models and programmes. In particular, it omits to acknowledge how a biopolitics of ageing relied on a downgrading of the status of fertility in determining the quantity and the 'quality of population'. This is somewhat in line with what Rose (2009) proposed to be the transformation of what I labelled the eugenic agencement in the 'politics of life itself': a shift from the classification of the eugenic value of persons to a valuation of the genetic and molecular components of ways of living. However, my view is that this relied on opening up death to biopower, just as it had been opened up by the disciplinary power of the projective gaze in pathological anatomy (Foucault, 1973). Death, and its relation to ageing, vitality and productivity, became a matter of concern for the 'ageing society'. This was the result of a series of epistemic and policy lock-in processes, whereby raising the efficiency and productivity of existing human capital became the only possible, imaginable pathway for managing the economics of ageing populations.

Chapter 5

Re-quantifying age?

Introduction

In the previous two chapters, I have examined how the 'ageing society' became a useful and meaningful descriptor for contemporary, uncertainty-laden, technological polities. This was divided in two parts. I first suggested that the 'ageing society' denoted a 'crisis' in the knowledge formats and practices that had underpinned the development of welfare states, and the creation of old age was a matter of administrative concern. This led, I argued, to framing uncertainty within the 'ageing society' around technoscientific promises geared to modify health and productivity through a set of converging tools that range from the molecular to the sociological. I then explored how this focus on health and productivity has been resolved within population science and policies since the late nineteenth century. Here, I argued that the relationship between health and longevity has become central to the problem of population because of how the concern with the 'quality of population' itself disqualified immigration and then fertility as solutions to the problem of demographic ageing.

There were, however, two key obstacles to this epistemic and political investment in what later became known as healthy ageing – and its re-evaluation of existing human capital within stable populations. One was the widespread belief that there was an inexorable inverse relationship between ageing and productivity or 'vitality'. As this relationship became a matter of debate at the turn of the 1980s, another hurdle arose, which related to the methods used to explore this relationship. How should ageing be measured, and what is its relationship with health and functionality? In this chapter, I explore why and how the issue of age mensuration became central to the constitution of the 'ageing society'. Doing this entails, first, understanding how chronological age (CA) – the number of years lived since birth – itself became a useful index for persons' functional capacity or health. This, in turn, requires seeing CA as both a form of measurement and an institutional format (Desrosières, 1991) – an agencement, to be exact. CA developed as the marker for a specific assemblage I earlier described as the 'insurance society', that is to say, for the modern *age-stratified* system of public rights and duties that includes taxes, schooling, the military draft and access to welfare rights such as pensions (Kohli, 1986).

From this perspective, Treas (2009) has detailed how the increased use of CA was linked to the requirement of precision, certainty and impartiality in a wide range of classification practices and decision-making procedures inherent to modern bureaucracies, binding epistemic norms within statistics with the information requisites of state administration in the latter part of the nineteenth century. As administrations extended their gaze and domains of concern, this bond worked to raise awareness of CA and its normative implications in European and North American societies (Chudacoff, 1989; Bytheway, 2011). But this was always a problematic agencement. As suggested in previous chapters, the use of CA as the standard for pension entitlement had been a controversial idea since the establishment of the first old age pension schemes such as that created in 1908 in Britain (Chapter 3). In this chapter, I first suggest how contradictory administrative rationalities used in the implementation of CA sustained a continued dynamic of critique and justification about the moral and epistemic worth of age measurement. I describe how such critique was concurrent with the establishment of gerontology as a field of knowledge in the mid years of the twentieth century, in its attempts to re-evaluate the worth of older people in society. In the third section, I explore how alternative forms of age measurement – emphasising functionality or health – emerged from this critical movement to integrate wider shifts in approaches to standardisation in late modernity, whereby measures and scales promise individualisation and 'personalization' of technologies or services (Busch, 2011). The exact sociotechnical configuration of such promises was, however, uncertain, and, in the final section of the chapter, I examine how one such proposal – the concept of biological age – became the object of a controversy that lasts to the present day. Through this analysis, it is possible to identify two diverging ways of enacting the relationship between normative ideals of the life course and methodological approaches to knowing and managing the relationship between ageing, health and illness – two approaches that we will come to recognise as fundamental to the patchwork of the 'ageing society', as we encounter them in subsequent chapters of this book.

Chronologising age

To analyse the assembling of the 'ageing society', in previous chapters, I underlined the role of the age pyramid in mediating – enabling calculations on – the relationship between the mechanisms of life and the economy that has become central to modern societies. To a certain extent, this analysis relied on 'blackboxing' the procedures through which age groups could be quantified. But what enables persons to be counted as members of an age group? Any answer to this question should begin with the understanding that days, months or years are not natural units of measurement of biological time. To understand how CA is able to index ageing processes in practice, it is useful to think, drawing on Desrosières (2008), that measurement of human characteristics is often deployed at the juncture between qualification and quantification. His argument is that quantification

– meaning putting in numerical form – is interdependent on a specific type of work that aims to construct conventions and classes across individual cases. In the case of CA, this means that equivalences are drawn between individuals of different genders, classes, localities, shapes and sizes on the basis of the number of years lived since birth. To draw this equivalence, Desrosières argues, requires investment – or durable commitment – in one 'coherent cognitive and political schema' that enables the qualification of persons or objects of concern (Desrosières, 1991: 214). Measurement enacts a particular repertoire of valuation, that is to say, it relates human characteristics or behaviours 'to some relevant good, the format of the good being highly variable' (Thévenot, 2006: 111; Moreira, 2005).

From this perspective, age measurement is a type of investment in form (Thévenot, 1984), a costly operation attempting to embed in codes, categories, standards or implements a stable relation between individuals' qualified characteristics and social institutions at different points in time. In her research into the institutionalisation of CA, Treas (2009) details the costly operation that underpinned CA being accepted as a stable measurement routinely collected by statistical agencies, firms and researchers of various disciplinary inclinations. Her argument is that there was an interdependent relation between the development and implementation of age-segregated administrative policies and the development of an apparatus of age collection and processing in statistics. In this process, the numerical precision that we normally see as an inherent quality of CA was, in reality, dependent upon complex and continuous work – the costly operation – of educating, policing and recording knowledge of age among the population (Treas, 2009: 74–76; see also Beaud and Prévost, 1994). This produced a positive feedback loop whereby CA as a product of fieldworkers' and statisticians' activities reinforced the capacity for administrations to formulate age-segregating policies, making it more imperative for citizens to develop age consciousness and knowledge, in turn easing the burden on later census workers, and facilitating the implementation of further policies based on the measurement of CA.

But what was CA measured for? What are the possible schemas of qualification under which it was – and still is – relevant to measure and quantify human ageing? Addressing this issue entails analysing the underpinnings of the link between the consolidation of modern administrative systems of distribution of rights and obligations and CA. Insurance societies are based on a mechanism of state-mediated solidarity whereby the vitality and productivity of the whole are managed by protecting against individually incurred risks. A key example of this are the old age pension programmes established in Europe and North America between the late nineteenth century and the 1940s. Their aim was twofold. On the one hand, they aimed to address the risks associated with later life (poverty, illness, isolation, etc.), sometimes measured through means testing and such like. On the other, as I have suggested, they sought to incentivise thrift and productiveness in working life, secured through contributions or 'industry tests'.

However, as we also saw, in implementing those policies, old age pensions also became controversially reliant on a minimum age, leading prominent reformers such as Beveridge to wonder if this age condition could trump industriousness (Beveridge, 1943: 96). The significance of this tension has been usefully analysed by Kohli:

> Chronological age is apparently a very good criterion for the rational organisation of public services and transfers. It renders the life course – and by that the passage of individuals through social systems – orderly and calculable. It is interesting to note however that there is an uneasy tension between the formal rationality of such procedures and the substantive rationality that they are supposed to provide. Chronological age is essentially an ascriptive criterion and thus at odds with the modern emphasis on universalism. [...] In a universalistic regime, it is normatively preferable to allocate right and duties by a criterion based on achievement, such as 'functional age'. Empirically on the other hand the implementation of such criterion is difficult and may even be self-defeating [...] replacement of chronological age by 'functional age' [...] would be very costly in several respects.
>
> (Kohli, 1986: 286–287)

Drawing on Weber's analysis of the conflict between formal and substantive rationalities (Weber, 1978: 225), Kohli suggests that CA brings forth an inconsistency between the application of the means for the effective administration of rights and duties and the ultimate goals for which this administrative system was designed. The latter should be linked to the application of criteria that qualify individuals in terms of achievement or need, and be ignorant of particularising characteristics. However, as Kohli explains, in the case of CA, the use of attainment or need requirements would be challenging to implement. There are two main reasons for this. One relates to the different age-related achievement criteria that would have to be deployed for different areas of state control and action. This would entail finding a different measure to qualify individuals for criminal responsibility and pensions, for example. Furthermore, such diversity would have to be ultimately validated within appropriate, equivalent epistemic and political schemas, themselves suitable for use within and across a modern universalistic state. Due to the complexity of developing and coordinating between different schemas, various policies and programmes have instead used CA as the criterion for qualification.

CA can thus be conceived as a compromise between rationalities or repertoires of valuation. But, with the expansion and consolidation of age-segregated programmes, this compromise has become increasingly unstable. One reason for this pertains to the impurity of CA as a convention, its belonging to two orders of worth. This compound quality facilitates critical engagement, making it easier to publicly denounce as a manufactured, artificial composite of different coherent epistemic and political schemas (Boltanski, 1990, 2011; Boltanski and Thévenot, 1999). This impurity has, in effect, powered continued and sustained critical engagement with CA from the very moment it became a staple, routine measurement.

There are, however, good reasons why, despite its problems, administrations use CA as a standard. CA is an 'empty' variable, that is to say that in and of itself CA does not provide information regarding the behaviour of the individual on which the measurement has been made. This means that CA has been used mainly in the form of a 'proxy variable' for the measurement of a variety of other criteria that the state or scientists were interested in: risk of unemployment, decreased functionality, etc. In this respect, CA has operated as a 'boundary object' (Star and Griesemer, 1989), supporting the activities of different 'social worlds' drawing on diverse repertoires, yet maintaining a common identity across them. While not fully satisfying the requirements of any specific use, this role has led to it being recurrently collected and processed by statisticians for a wide array of purposes. The main consequence of this is that age measurement – and by extension the representations of the 'ageing society' that it enables – has become 'locked-in' CA. Over time, the incumbency of CA and its plasticity have worked together to maximise its practical advantages, while raising the bar for the development of any alternatives. Thus, seeking to replace CA with another form of age measurement, as Kohli has noted, would be 'very costly' because in order 'to establish other links and new translations you would first need to undo those already in existence by mobilising and enrolling new alliances' (Callon, 1991: 152).

This paradoxical relation between the contradictory foundations of CA as a governing tool and the enormous cognitive and political costs associated with replacing it has shaped the way in which it has become the object of a dynamic of critique and justification since the turn of the twentieth century. This has meant that a widespread and trusting institutional reliance on CA to appropriately ascertain age-specific norms, values and expectations has been replaced by epistemic and normative uncertainty not only about CA as a 'variable' and/or marker for social and political rights and duties, but also about any proposed alternatives to CA. In the sociology of ageing, this process is usually described as the de-standardisation of the life course: an increase in the overall heterogeneity – individualisation – of transitions to adulthood and retirement (Beck, 2001; Brückner and Mayer, 2005; Kohli, 2007). From our perspective, however, this de-standardisation process is heavily reliant on the weighty uncertainties that have impended on the epistemic infrastructure – the forms of age measurement – supporting individuals' and organisations' decisions about life course trajectories. Star and Bowker usefully describe the strain between these levels as 'torque', whereby classificatory systems and individual trajectories mismatch and readjust (Bowker and Star, 1999: 223). In the case of age measurement, as we will see below, it is not only that discrepancies appear between individual age identities and age categorisation systems, but also that inconsistencies between different alternative age measurements and their valuation repertoires are brought to bear. Identities might become torn within and across age classification systems.

Problematising chronological age

In his opening remarks at the 1954 CIBA Foundation Colloquium on Ageing, Professor R. E. Tunbridge, OBE, asked the following questions:

> Is ageing a chronological term, merely reflecting the passage of years, and if so, what years, or are the public right in assuming, as they generally do, that ageing is synonymous with senescence and/or decay? The concept of the elixir of life [...] has long served as a tremendous stimulus to mankind, to higher flights of imagination or sometimes to derision. We shall not dwell upon these fantasies, nor shall we deal with that other very important aspect of the problem, what one might call the political, economic and social aspects of ageing, of which we as citizens cannot be unaware.
>
> (Tunbridge, 1955: 1)

Tunbridge was well placed to ask these questions. Not only was he by then an internationally respected physiologist for his work on insulin processing and collagen tissue, he also had a key role in coordinating the date of the colloquium with the Third International Gerontology Congress. For this reason, in the audience there were some of the most prominent researchers in the emerging field of gerontology. These included: Sir Frederic Bartlett, of the Cambridge Experimental Psychology Laboratory; Peter Medawar, then at University College London, and later to receive a Nobel Prize in physiology; Nathan Shock, Director of the US Public Health Service Section on Gerontology; and Edmund Cowdry, founder of the International Association of Gerontology, and organiser of the 'problems of ageing' symposium seen as the origin of modern research on ageing (Park, 2016).

From this position, he posed the most pressing question, of whether ageing was 'merely' a chronological measurement or related to functional and physiological decline. This question, Tunbridge argued, allied the experts in the room to the concerns of the wider population, who, while sometimes drawn by fantasies of immortality, tasked scientists to address one of the most important social and political issues of the time: the 'ageing society'. By making explicit the connection between science and the unyielding questions of the 'ageing society', Tunbridge was also implying that uncritically accepting the chronological understanding of ageing was no longer politically imaginable. To respond to the concerns of the public, it was necessary to explore how years lived since birth related to 'senescence and decay', if at all. With this reference to the wider public, Tunbridge was, in effect, rhetorically acknowledging that the experts in the room had a unique opportunity to shape the public agenda on ageing. It was the time to bring to the public sphere questions about the function and role of CA in administering populations. In this, he was backed by five decades of expert interrogation of CA.

The origins of this examination can be traced back to the consolidation of the sciences of growth in the US at the beginning of the twentieth century. In that period, as we have explored in previous chapters, progressive reforms of the American nation combined with growing anxieties about modernisation to induce the expansion of privately and publicly funded research on child development (Smuts, 2008; Prescott, 2004). These aimed to replace the previous concern with poor and delinquent children with a scientific understanding of the 'normal' child. Between the mid 1920s and the early 1930s, researchers in the emerging field of child development established a variety of studies aimed at examining 'growth' by means of serial observations of selected children, collecting both biometrical data and psychosocial assessments (e.g. the Harvard Growth Study; the Oakland Growth Study).

The concepts and ideals of the child development research movement can be seen as embodied in the figure of Lawrence Frank, one of the key planners of the movement. As discussed in Chapter 2, Frank, as a social scientist, was typically troubled by modernisation and what he saw as the resulting, growing disjunction between habitual human behaviours and industrial, technological culture. As the manager of the Laura Spelman Rockefeller Memorial Fund (LSRMF), and as an editorial board member of *Growth* – the key journal in the field of child development – Frank advanced the proposal that the understanding of processes of normal physiological and psychological development was key to the design of beneficent social institutions and the management of individual behaviour – of the 'design for living', to use Frank's words (Bryson 1998: 410). As we also saw in Chapter 2, as the federal government became entangled in debates concerning the future Social Security Act (1935), Frank was again called upon to create an equivalent platform for ageing research, as vice-president of the Josiah Macy Foundation (Achenbaum, 1995: 76–79).

This resonated with the vision proposed by Edmund Cowdry, the main scientific instigator of the role of the Macy Foundation in sponsoring research into the 'problems of the elderly', leading to them both organising the symposium later published as *Problems of Aging* (Cowdry, 1939). As Park (2016) has documented, Cowdry, a cytologist working at Washington University, had been concerned with understanding the differences between physiological ageing processes at the level of the tissue or organ. Drawing on the ideas of Nobel Prize-winning physiologist and eugenist Alexis Carrel, Cowdry proposed that the rate of ageing in tissues was determined by their surrounding environment of nutrients, regardless of the organism's chronological age. This explained why,

> ....the burden of years is not evenly felt by blood vessels of all sorts. In addition to such local differences in susceptibility remarkable differences in speed of operation of the ageing processes are noted.
>
> (Cowdry, 1939: 665)

By disentangling the 'operation of the ageing processes' from a singular measurement of time, Cowdry detailed the case against CA from a biological, physiological perspective. In particular, he questioned the idea that there was an alignment between the calendar and the various 'speeds' at which different organs develop and decay. If welfare institutions, professionals and experts were to address the 'problems of ageing', which appeared to be the aim of the Social Security Act, it was necessary to do away with the notion that years lived since birth could index physiological and functional status. The principal reason for why this idea was so central to Cowdry's work – and would become the core of the physiology of ageing – was because it was fully aligned with contemporaneous medical and scientific ideas and practices regarding growth and ageing, whereby 'the old temporal delimiters of birth and death have become blurred and natural time has been increasingly stripped from the body' (Armstrong, 2000: 258).

Frank's role was also crucial in helping establish another key figure in modern gerontology, Nathan Shock (see above). Shock had begun his academic career in the mid 1930s as a researcher in the Oakland Growth Study, a study in which Frank had invested great hope towards the end of his tenure at the LSRMF. Focusing on physiological changes in adolescence, Shock had been able to establish that the onset of a physiological event – menarche – was more significant than CA in determining developmental change. It was thus no coincidence that, when advising the federal government on the establishment of an intramural gerontology programme at the National Institutes of Health, Frank recommended Shock as the new chief of the Gerontology Unit, when the first chief, Edward Stieglitz, resigned for personal reasons (Park, 2016).

There, Shock was able to establish, during the 1940s, a programme of research that drew on his experience in measuring 'physiological age', first using institutionalised elderly subjects from the Baltimore Department of Public Welfare and later when establishing the Baltimore Longitudinal Study of Aging (see Chapter 6). In this, Shock could be seen as the spokesperson for an approach to ageing research that focused on individuals 'from different points of view and at different levels of organization' (Shock in Wolstenholme and Cameron, 1955: 244), rather than solely on organ systems, as Cowdry would suggest, or as members of populations, as had been proposed by Medawar (1946). This angle, Shock suggested, provided the strongest counterpoint to CA, and the assumptions about old age embodied therein. This was a necessary epistemic platform for policies and programmes handling, as he later put it, 'older persons as individuals, not as a class, and their wide differences in needs, desires and capacities' (NWS: 21: 'Progress in the field of gerontology', 13 July 1956).

In psychology, the critique of CA is perhaps most visible in the work of Ross McFarland. Having worked in the Cambridge Psychological Laboratory with Frederic Bartlett, his work on pilots' cognitive performance and aeroplane design was decisively shaped by his collaboration with Henderson, Mayo and others at the Harvard Fatigue Laboratory (see Chapter 7). His concern with the inadequacies of CA in assessing function was sparked,

> [in] WWII when it became necessary to employ a large number of retired older workers in the war industries, especially in the aircraft manufacturing

companies of Southern California. At that time a study was made, The Older Worker in Industry, reporting that older workers, if properly placed, could function effectively. In fact, they had greater stability on the job, fewer accidents and less time lost from work as did younger workers (McFarland, 1943). The investigation showed it was unfair to judge workers in terms of their chronological age.

(McFarland, 1973: 1)

Recalling a situation where the age-segregating policy of retirement had been suspended, McFarland reported on the inadequacy of CA to index function, leading to an inadequate organisation of labour. Strikingly, McFarland explicitly framed this discrepancy in normative terms, in that it prevented persons from being recognised, and paid, for their abilities. For McFarland, chronological ageing 'cannot be arbitrarily evaluated as good or bad, but rather that [it] must be clearly understood in relation to the demands of specific jobs or employment possibilities' (McFarland, 1956: 21; also McFarland and O'Doherty, 1959). This belief in an arbitrary, unfounded, unjustified evaluation of people 'as good or bad' according to CA had also motivated the Nuffield Foundation to support the work of A. T. Welford and Bartlett at the Research Unit into Problems of Ageing linked to Cambridge Laboratory from 1946 (see Chapter 7). This research suggested that most of the issues arising from older people in industry had come about as a result of changes in technology, a generational effect that had become crystallised in CA-based retirement policies after the war (Thane, 1989). This further denounced the historical contingencies upon which CA measurement was based.

Frank's role in this critical engagement was also evident in the social sciences. In appointing Robert Havighurst to the Committee of Human Development at Chicago, while it was still mainly concerned with psychological child development, Frank facilitated the involvement of social scientists in the design of educational and research programmes in social gerontology in the late 1940s, which proved decisive for the growth of the sub-discipline. This led to the establishment of the Kansas City Study of Adult Life where, drawing on a Chicago 'community based' style of social science research, social gerontology's first 'social science laboratory' was established (Achenbaum, 1995: 106). Two important sociological approaches to ageing came from the study. One was the theory of disengagement (Cumming and Henry, 1961), which, echoing somewhat the wishes of Frank, argued that the 'normal curve' of development and senescence was paralleled by a decoupling of moral obligations of adult life and personality of older people. The other, later embodied in the figure of Bernice Neugarten, focused instead on how age norms interacted with personality and biological changes to shape behaviour (Neugarten, 1964).

Of key significance in the development of this approach, particularly for our purposes, was Neugarten's study on the meaning of menopause for middle-aged women. Contrary to physiological models of development and senescence, the study's findings suggested that menopause was 'not necessarily an important event in understanding the psychology of middle aged women' (Neugarten, 1968: 142). Her argument was that, in later life, age norms and 'age status systems' were more important to understand behaviour and personality than biological events. However, when investigating age norms, Neugarten found that at the phase in life when there was most variation across individuals, there was also 'an increase in the extent to which respondents ascribe importance to age norms and place constraints upon adult behaviour in terms of age appropriateness' (Neugarten *et al.*, 1965: 715), a cultural rigidity that successive age-segregating retirement policies in the US had re-enforced.

Later on, Neugarten was able to identify a renewed fluidity in age grading systems, prompting her to hypothesise that we might be observing the birth of,

> an age-irrelevant society in which arbitrary constraints based on chronological age are removed and in which all individuals, whether they are young or old, have opportunities consonant with their needs, desires and abilities.
>
> (Neugarten, 1974: 197)

Such a vision of an 'age-irrelevant society' where the distribution of rights and responsibilities hinges on ability and need, while normatively not radically different from those proposed by Cowdry or McFarland, is enriched by an empirical attention to different practical, socially embedded ways of growing old. This contributed to questioning the inevitability of contemporaneous age grading systems – its arbitrary constraints – and to denaturalising the life course within social science through the consolidation of the institutional paradigm, of which Kohli's 1986 paper used above is a prime example (Dannefer and Settersten, 2010).

In Britain, the creation of the National Health Service (NHS) and the National Insurance Act sparked a renewed interest in 'those elderly sick with social and economic problems' (Lord Almuree in Martin, 1995: 460), an interest played out in initiatives such as the Nuffield Foundation's Survey Committee on the Problems of Ageing and the Care of People and its sponsoring of a research programme dedicated to old age. It was through this network that British social gerontology first emerged in the form of Townsend's *The Family Life of Old People* (1957). For this, Townsend was located at the Institute of Community Studies, a research centre established by Michael Young to escape from the constraints of British academia and study the lives of ordinary citizens. In this aim, he had drawn on the ideas of Edward Shils, a Chicago sociologist who supervised Young while serving on a teaching post in Britain. Shils was interested in challenging the view, much enforced by his former collaborator Parsons, that 'modern society [is a] Gesellschaft [...] lacking any integrative forces other than interest and coercion', and proposing instead that 'it is held together by an infinity of personal attachments and moral obligations' (Shils, 1957: 345; see also Bulmer, 1996).

Similarly, Townsend collected interviews with older people living in Bethnal Green to demonstrate that the extended family and neighbourhood relations had survived and transformed with the establishment of the welfare state. This led to the claim, developed in subsequent studies, that usage of pensions and care was significantly determined by the family structure of the older person rather than by age or older people's physical or psychological 'dependence'. As he recounted,

> In 1954, when Michael Young started the Institute of Community Studies, I had the opportunity to interview old people in Bethnal Green. What struck me hardest was the extraordinary diversity between people of similar age living in the same locality. It was deeply puzzling. All the stereotypes in one's mind had to be taken to bits. By the time I had finished counting exceptions to the politician's traditional picture of Darby and Joan living on the old age pension, there was nobody left.
>
> (Townsend, 2009: 153)

This disjunction between the 'stereotype' and the diversity of living conditions and needs struck Townsend as deeply contrary to the vision of an equal, just society advocated by welfare state theorists – e.g. Beveridge, or Titmuss. Such questioning of the fundamental fairness of welfare policies, and of retirement policies in particular, emphasised social and psychological needs and abilities of persons rather than the presumed need to alleviate the 'economic burden' of older people in society. CA was an obstacle to the ideal of fairness espoused by social policy researchers like Townsend, because it lacked the capacity to distinguish and respond to the 'extraordinary diversity between people of similar age'.

Reimagining age measurement

From the 1940s onwards, across biological, behavioural and social sciences, a mounting critique was directed at the foundations of CA. The critique highlighted the epistemic inadequacies – the artificiality – of the age measurement system where equivalences were drawn between individuals with a wide diversity of physiological, psychological and sociological characteristics. In this process, CA became increasingly seen as an arbitrary age standard. The critique also denounced the way in which CA deployed an unfair treatment of persons by medical, work and welfare institutions, classifying people as 'good or bad' on the basis of years lived alone. Thus, in proposing alternatives, age measurement was instead seen as underpinned by the mobilisation of expert knowledge to distinguish, or differentiate between, persons previously categorised as equivalent.

Substitute age standards were, in this respect, part of the wider shift in standardisation in late modernity, whereby measures and scales promise individualisation and 'personalisation' of technologies or services. Their purpose is, as endocrinologist Harry Benjamin put it in explaining the goal of coining the concept of biological age, to develop 'for the individual what our actuaries and biostatisticians have figured out so ingeniously for groups' (Benjamin, 1947: 226). As Busch (2011) and Epstein (2008) have suggested, the goal of such standards is not to develop a universalistic measure, applicable to all, but also not to rely solely on individualistic assessments. Instead, experts focus on the conception, validation and implementation of standards which purport to identify the combination of unique characteristics of persons that are relevant to a specific market, service, technology, or type of work. What characteristics are considered relevant for each setting is, however, a function of the repertoire of valuation brought to bear in designing alternative measurements.

It is possible to identify four types of repertoires that have been drawn on to propose a new form of age measurement since the 1930s, depending on how they combine the focus of the measurement with the form of good that it is justified by. On the one hand, qualification can range from a focus on behaviour in the concepts related to functional age, to an emphasis on somatic qualities, on measurement clustered around the idea of biological age (see Salthouse, 1986). On the other, measurements either aim to maximise older people's economic and social productivity – usually associated with proposals that wish to realign instrumental and substantive rationality in age measurement – or seek to provide the means through which persons can achieve their personal life goals, desires and ambitions throughout the life course, emphasising the value of individuals' unique personal characteristics, such as wisdom, inspiration, extraversion or creativity.

Propositions which focus on somatic qualities and emphasise efficiency are typically associated with the concept of biological age, as in Benjamin's paper referred to above. He argued that these forms of measurement would be important for one main purpose: knowledge of individual BA could assist in monitoring and assessing the effects of healthy living, medical interventions or 'gerontotherapy, which indicates the treatment of the aging process as such'. (Benjamin, 1947: 225). It was also from this perspective that Alex Comfort approached this subject. Nowadays mostly known as the author of *The Joy of Sex*, Comfort was a key figure in the consolidation of biological gerontology, having written the first textbook on the subject (Comfort, 1956). A student and follower of Medawar's (1952) evolutionary understating of ageing, Comfort became known in the 1950s and 1960s for promoting an approach to the measurement of ageing hinging on the effects of Gompertz' 'force of mortality' upon survival curves (Comfort, 1956: 22–44; see Chapter 3). His need to find an application for the biology of ageing within the problem-driven research policy environment of the 1960s, combined with a renewed confidence in new pharmacological approaches to ageing – particularly anti-oxidants – led him to the view that while,

> for a proper account of ageing it has so far been necessary to insist on the force of mortality as the sole generally applicable criterion, [now a] new attempt to work out battery tests of human physiological age is overdue. It

is justified by experimental necessity. Agents are known which seem to prolong the life-span of rats and mice [...]. It is highly probable that some of these would affect human life-span if they could be tested briefly and ethically.

(Comfort, 1972: 101)

Comfort's 'battery test' was explicitly justified by changes in medical technology. According to him, agents 'which seem to prolong the life-span' entail the development of new metrics for both practical and ethical reasons. Such metrics should be able to detect, with minimal risk for participants, proximate changes in underlying biological processes before their ultimate, temporally distant effect – the postponement of death – can be measured. In a proposition that anticipates much of what is now understood to be the function of 'biomarkers of ageing' (see below), Comfort articulates a vision for the biological measure of ageing that embodies the technological promise of ageing research. In this evaluative role for the measurement of biological age, the metric assesses the effects of presumed anti-ageing therapies and facilitates their effective implementation through the bioclinical management of individuals.

Measurements that assess function and behaviour in relation to a valuation of efficiency are more diverse than the previous type, as they span from strict measurement of specific functions to generalised assessments of social functioning (e.g. Lawton and Brody, 1969; see Chapter 8). Their goal is to produce a measurement of individual functional abilities (or disabilities), and imagine a corresponding articulation of these with the tasks the individual might be asked to perform, or the services or goods s/he might be entitled to. As was discussed above, a central concern within this group of substitute measurements is the possible nefarious effect of technology on work (also Chapter 7). An indication of this apprehension is revealed in a review of measures of age conducted for the WHO by François Bourlière in the late 1960s. In this, he argued that,

It is evident that the 'wear and tear' of existence does not show in the same way and at the same rate for all of us. [...] This entails a series of problems of concern not only for the physician and the psychologist but also to the sociologist and the economist. In particular, the diversity we display in the pattern of ageing has implications for professional life. Every possible effort should be made to adapt the type of work done to the changing capacities of the individual.

(Bourlière, 1970: 9)

Bourlière's proposal was related to that of McFarland discussed in the previous section, but was underpinned by a narrative of the effects of modernisation in French society. In the 1960s, Bourlière led a series of studies at the Claude Bernard Gerontological Centre that compared ageing processes – measured physiologically and behaviourally – in 'traditional' and 'modern' occupations in France (e.g. Bourlière *et al.*, 1966). These studies revealed that while traditional jobs were associated with faster decline in function, individuals in modern occupations were more likely to become inactive due to changing work conditions and retirement policies. This not only had an effect in terms of unused resources for the economy, but also on how such individuals were valued in a work-centric society. Bourlière thus proposed that new metrics were necessary to adjust the 'changing capacities of the individual' to the 'type of work' performed by him/her. This would give rise to work and welfare institutions that would use such technical means to regularly assess capacities and requirements/needs to achieve efficient deployment of resources.

The third type of measurement of age concentrates on function but as related to the pursuit of an ideal of individual uniqueness. This seemingly paradoxical valuation is possible to articulate where the 'common good' is seen as best served in supporting individual creativity and 'genius', and attribution of value becomes based on the quality of subjective experiences (Boltanski and Thévenot, 2006: 98–106). Proposals in this quadrant focus on the views of self and others on their position on the life course, and on the ambitions and desires that were expressed in Neugarten's vision of the 'age-irrelevant society', or in Laslett's concept of 'subjective age', where personal achievement and experience are emphasised in detriment of public, social roles (Laslett, 1989: 193–195). The best examples to epitomise such an approach in terms of age measurement are the use of reminiscence and life review instruments (Settersten and Mayer, 1997: 249–251). Interestingly, a key figure in the development of these implements was Robert Butler, the American psychiatrist usually credited with having coined the term 'ageism' (Butler, 1969), and the first director of the US NIA between 1975 and 1982.

Butler suggested that reminiscence was a normal occurrence in ageing and that the life review could be deployed as an instrument to understand the evolution of 'personal characteristics that seemed to be associated with age, such as candour, serenity, and wisdom' (Butler, 1963a: 67). Butler contrasted this approach to ones where,

> many manifestations associated with aging per se reflect instead medical illness, personality variables, and social-cultural effects. [...] Intensive studies, involving frequent contact over considerable periods of time, based upon the growing personal relationship between the investigator and the older person, would contribute to our understanding of the subjective experience of aging [...]. If we can get behind the façade of chronological aging we open up the possibility of modification through both prevention and treatment.
>
> (Butler, 1963b: 721)

Taking the model of the psychotherapeutic relationship, Butler suggested that individual experience and prospects about the ageing process could be understood through and shaped by the life review instrument. The relationship and the tool were seen as fundamental in providing meaning to 'the possibility of modification' of experience but only if combined with other conventional measures. As Butler admitted, characteristics such as wisdom or candour were 'elusive concepts [...], difficult to demonstrate [and] even harder to measure' (Butler, 1982a: 35). Because of this fuzziness, instruments such as the life review alluded to – rather than represented – a subjective, individualised herme-neutic of experience. While seen to escape the usual requirements of precision and objectification, the deployment of these subjective qualities of individuals worked against the negative, medicalised view of ageing, providing a normative scaffold on which to support one's own construct of the life course.

Proposals of the last type link the identification of somatic qualities with enabling exceptional longevity. In this approach, centenarians and other long-living individuals are conceived as 'models' of ageing well. Measurement of chronologi-cal years is not used to index functional capacity but instead to defy the statistical norms and normative expectations regarding old age. Exceptional longevity becomes a measure of healthy, successful ageing rather than simply years lived; as the investigators of the New England Centenarian Study put it, 'the older you get, the healthier you've been' (www.bumc.bu.edu/centenarian/overview/). This focus on exceptional longevity continues a tradition of research on the 'biological unique-ness of long lived individuals' that had been first articulated by Nathan Shock and his colleagues in the Baltimore Longitudinal Study of Aging (see Chapter 6). Such persons are seen as 'model organisms', and are valued in research for being out-liers, and not typical or average. They are also seen as moral examples, embodying the markers of the 'good life', longevity being associated with 'good personal qual-ities' such as extraversion and low neuroticism.

The vision within this approach is that by studying, measuring and learning from the biology of these exceptional individuals, we might be able to extend the potential for longevity to significant portions of the population. In clarifying the political underpinnings of this proposal, the work of Alex Comfort is of use again. Differentiating between the purposes of mainstream medicine, and those of the biology of ageing, he argued that,

> Insofar as biology is more than a branch of idle curiosity, its assignment in the study of old age is to devise if possible means of keeping human beings alive in active health for a longer time than would normally be the case – in other words to prolong individual life. People now rightly look to science to provide the practical realisation of perennial human wishes [and] medicine can afford to treat protests based upon an interested misreading of biology of human societies [as] a compound of illiberal opinion and bad science.
>
> (Comfort, 1964: 270–271)

Refuting a neo-Malthusian reading of the work of August Weismann that would emplace a moral worth on dying for the benefit of the species' survival

(Moreira and Palladino, 2012), Comfort proposed that biology of ageing was most adequate to fulfil a libertarian agenda. An anarchist himself, Comfort was deeply concerned with how social norms and political prejudices about older people imposed unjustifiable burdens on their individual freedom (Comfort, 1977). Removal of the restraints associated with CA could only be brought about by 'a compound' of libertarian movement and good science, which he associated with a precise and robust measurement of senescence. As biologist of ageing Tom Kirkwood put it in his BBC Reith Lecture in 1999, this entails a challenge to 'look in radically new ways at the maintenance of health and quality of life of older people [and to] imagine a world in which the first thing the doctor asks is not your date of birth' (Kirkwood, 1999: 215)

The proliferation of these four types of age measurement was supported by a shift in the institutional infrastructure of standardisation processes, linked to major transformations on the function and power of the state in governing polities. This has meant that whereas before the state had a central role in the production and validation of standards, in the neoliberal era, these activities have become decentralised. As Desrosières (2008: 12) has argued, the polycentric multiplication of networked centres of decision within globalised, 'financialised' markets has led to a proliferation of standards-making agencies and institutions. In age measure, since the late 1960s, an increasing variety of organisations have proposed substitute age standards: universities, pharmaceutical companies, biotechnology firms, health insurers, occupational health agencies, charities, social care providers, etc. Companies have been set up whose only objective is the provision of age measurement.

This expansion and intensification of research on 'personalised age' has not, however, built a growing consensus, with papers instead proposing ever different 'markers', techniques of measurement, or approaches to statistical calculation of battery tests. As the American Federation of Aging Research put it as recently as 2011, '[w]hile there are several candidates for markers of ageing, none have so far proven a true measure of the underlying aging process.' (AFAR, 2011: 2). Despite the robust valuation repertoires, there has been continued debate and uncertainty about the purpose, accuracy, reliability, practicality and safety of alternative age standards. Why is this so? Why are new age standards difficult to validate and implement? This is the focus of the next section.

The 'biological age' controversy

In the first section of this chapter, I have argued that the process of substitution of CA for another form of age measurement would be constrained by the simultaneous contradictory nature of CA and its pervasiveness. While the moral and epistemic impurity of CA makes it an easy target for critique, its 'emptiness' allows for a myriad of combinations and normative purposes to be embodied in it, resulting in an institutional and material entrenchment of CA in our daily lives. This enables a dynamic of critique and justification whereby the pros and cons of replacing CA are explored. Nowhere is this dynamic more evident than in the controversy about 'biological age', the first type of age measurement explored in the last section. This involves tools that aim to relate the measurement of somatic qualities to efficiency, mostly conceived in terms of a functionalist understanding of health.

For this reason, they have been specifically linked to the research agenda around health, activity and technology, and the corresponding technoscientific promises of the 'ageing society', discussed in Chapter 3. In this respect, they can be said to have included – collapsed onto – some of the techniques and objectives of the more 'physiological' kind of functional age measurement, which we will explore in Chapter 7. Here, health – and its measurement – mediates between work efficiency and the value of technology. This mediating, transforming role was, in this context, particularly invested in the concept of 'biomarker of ageing', understood as 'a biological parameter of an organism that either alone or in some multivariate composite will, in the absence of disease, better predict functional capability at some late age, than will chronological age' (Baker and Sprott, 1988: 28). Indeed, the need for new measures of individual biological ageing was intimately linked to the socio-epistemic challenges brought about by the social security crisis. As Butler put it as Director of the NIA,

> [t]here are both scientific and socioeconomic imperatives for developing biological makers of aging. The scientific imperatives derive from [...] the possibility that certain age-related phenomena [...] may be controlled through intervention, [the testing of which] is dependent upon accurate measures of biological aging. The socioeconomic imperative [stems] from economic perturbations that have threatened the integrity of the Social Security System, [and which have motivated] proposals to increase the age of social security eligibility. [...] Because of the increased age of the workforce and conflicts over retirement age [...], we must be able to assess properly the impact of aging on human performance.
>
> (Butler, 1982b: vi)

Bringing together the technologically focused justification deployed in the measurement of the type 1 analysed in the previous section and those closer to the second type of age standards, Butler reaffirms the necessity of using 'accurate measures of biological aging' as a means to value and enhance existing human capital. This collapsing of the two types of age measurement, while representing a reduction of the space of possibilities for reimagining ageing, further evidences the need to strengthen the assemblage around the pivotal role of health in the ageing society. Doing away with CA was, in this regard, heavily linked to an alternative vision of the relationship between knowledge making and the management populations across the life course. Butler's defence of the 'imperatives for developing biological makers of aging' was also a means of shoring up this technoscientific imaginary. To understand the agonistic character of Butler's statement, it is necessary to go back to one of the first conferences organised by the NIA, the 1977 Epidemiology of Aging Conference.

Aiming to 'derive new knowledge to advance our understanding of the underlying causes of the aging process and help us separate disease from aging' (Butler, 1980: 4), the conference was an important occasion because it marked the distinctive identity of the NIA in the federally funded research infrastructure. Bringing together biologists, social scientists, clinicians and statisticians, the conference rearticulated the multidisciplinary vision of gerontology which Cowdry and Frank had fostered in the 1930s (see above). This approach was necessary because of how the NIA linked the challenges of the ageing society with the need for a new approach to knowledge making. Thus while the conference was specifically looking at epidemiology, its participants ranged from those used to working within laboratory settings – e.g. Richard Adelman, a biochemist from Temple University – to those concerned with studying ageing, illness and health in community-dwelling populations, such as William Kannel, one of the investigators in the Framingham Heart Study.

The first session of the conference, chaired by Samuel Greenhouse – previous head of the Division of Statistical Methods at the US Public Health Service – was concerned with 'attempt[ing] to formulate a definition of aging other than that of chronological age' (Haynes and Feinleib, 1980a). Two important contributions were included in this session. In the first, Adelman revised current experimental work on biological parameters of ageing, concluding that epidemiological work on BA should go beyond observation and record 'altered physiological response to hormones, drugs, antigens, exercise and other conditions of stress' (Adelman in Haynes and Feinleib, 1980b: 13). This was an endorsement of Comfort's techno-promissory vision for the development and validation of biological age measurement.

The second key presentation of the conference was provided by Drs Paul Costa and Robert McCrae, two psychologists then working at the intramural research institute of the NIA in Baltimore. Entitled 'Functional Age: Conceptual and Empirical Critique' the presentation reviewed the existing studies focusing on individualised age measurement from Benjamin onwards, to conclude that they represented 'a prescientific approach which showed considerable promise [...] but which would be best abandoned' (Costa and McCrae, 1980: 27). Their critique of somatic and functional age measurements hinged on three lines of reasoning. First, that it was irrational to want to devise a single or composite metric of ageing when it was known since Cowdry's work that different physiological systems age at different rates (see above). Second, that individuals' rate of ageing was unlikely to be constant through their life span. Third, that using a method which relied on regression to the mean contradicted the very aim of constructing individualised age measurement.

In relation to this last aspect, it is important to note that models of biological age calculation have relied on multiple regression techniques since Benjamin's (1947) original paper. This had been reinforced in Murray's proposal, drawing on the advice of the eminent statistician Ronald A. Fisher, that 'a method that would in some way combine various physiologic functions would give a more useful assessment of the individual' than single measurements (Murray, 1951: 120). To do this, he suggested research should use the multiple regression method to assess the differences between CA and BA of selected age groups. Mindful of the already established consensus that different physiological systems would age at different rates, Murray envisaged his metric as an average mean – a statistical approximation of the real ageing rate of the individual (see Chapter 6). For Costa and McCrae, however, it was not only inappropriate to use the same test battery for different age groups but also to rely on regression to the mean as a means to validate individualised age measurement. Further, they argued, using the difference between BA and CA as the criterion of validation of the former rendered the latter measurement somewhat useless, if the intention was to replace CA as a measure of age (Ingram, 1988).

Because of these methodological problems, BA could not, in their view, replace CA as an individualised assessment of individuals' capacity to work or of the effects of age-retarding interventions (see also Ludwig and Smoke, 1980). They concluded:

> Chronological age is indeed a dummy variable, and research should attempt to replace it in every case with an account of the real aetiology of age-related changes. The substitution of a new dummy variable, functional age, will not be a step forward on this process; indeed, it would be a step back.
>
> (Costa and McCrae, 1980: 45)

Their attack aimed at the very core of the sociotechnical imaginary associated with BA. Its promise was that BA would enable institutional and technological change underpinning older people's health and socio-economic participation. Costa and McCrae suggested that it could not deliver this undertaking. Instead, they argued that the current CA-based system had many advantages. One key advantage related to the universalistic nature of CA. This made 'chronological age [...] completely democratic, possessed equally by rich and poor, man and woman, healthy and sick' (Costa and McCrae, 1980: 44; see also Kohli, 1986). The invocation of the democratic political ideal upon which CA collection and calculation were based aligned Costa and McCrae's critique with a defence of the modern administrative apparatus, of the 'insurance society'. Theirs was a 'reformist' approach to the problems that had been shown to pertain to the use of CA in calculating life expectancy and relating it to 'vitality'. A second main advantage was related to this need to address the relationship between mortality and morbidity, as analysed in the last chapter, in that Costa and McCrae suggested that it should be possible to build 'for every case' profiles of 'real etiology of age-related changes' while using CA. In other words, CA was key not only for addressing the challenges of the ageing society but also for implementing the personalised approaches to health technology that alternative measures most promised.

The significance of Costa and McCrae's paper, evidenced by repeated print-ings of the paper and the high number of citations, is that it turned the tables. Whereas proponents of BA had been supported by a critical consensus around the inadequacies of CA, their paper questioned whether the solution proposed had any advantages in relation to the current CA-based system. The technoscien-tific promises linked to BA, they argued, could be delivered with CA. As Green-house put it in the discussion following the presentation, 'the onus [was now] on the other side' (Greenhouse in Haynes and Feinleib, 1980b: 47).

Thus, from the early 1980s onwards, starting with the 1981 Conference on Biological Markers of Aging referred to above, a series of concerted pro-grammes of action – workshops, funding initiatives, consortia, bringing together an extended collective of researchers, clinicians, public and increasingly private funding agencies, mainly in the USA but also in Europe and Asia – were organ-ised to build research capacity in this area. They can be seen as addressing, sometimes explicitly, the methodological issues raised by the 1977 presentation. Among these attempts to move BA from a 'prescientific' to a scientific stage of development, the NIA Biomarkers of Aging Funding Initiative between 1988 and 1998 is perhaps the most important turning point.

Supported on the back of raising political interest on the effectiveness of age-retarding interventions (interview with Richard Sprott, 10 December 2012), the programme aimed to develop a panel of biomarkers of ageing by assessing the effects of caloric restriction on genetically homogeneous strains of rats and mice. Deploying the conventions and instruments of laboratory physiological research, the programme leaders had hoped to find statistically significant differences between ad-hoc and restricted diets on particular measures of ageing. This was a challenging task involving the rearing, circulation and testing of a varied array of animals. By 1997, it had become clear that a panel of biomarkers was not going to be obtained (Sprott, 1999). This was a momentous setback for BA pro-ponents, because of how they had linked this programme to the promises of BA measurement. It was the more problematic because the reasons behind its failure could be clearly linked to the methodological critique enunciated by Costa and McCrae – and others – almost 20 years earlier.

Already at the beginning of the research programme, McClearn (1989) had warned that the protocols hinged on two problematic assumptions. The first was that there was a real rate of ageing which could be captured by a com-posite measure of different cellular systems and processes. The second was that this measure could be used for both constant and variable rates of senes-cence across the lifespan. These, in effect, reiterated part of the Costa and McCrae 1977 critique (Costa and McCrae, 1980). In addition, during the implementation of the programme, it became clear to interested, expert obser-vers that the research had not been adequately powered (interview with Richard Miller, 23 January 2013). This meant that it would not be able to find a significant variable of group of measures to explain the variations found in the population.

But the most poignant critique addressed the institutional, normative assumptions of the biomarkers of ageing programme. The fact that it was articulated by someone who just ten years earlier could be considered to be a proponent of BA, Richard Adelman, made it all the more piercing. Adelman was of the view that,

> history will regard research related to the identification of such biomarkers of aging as no more than of fleeting significance. Sophisticated research entails so much more than pure empiricism. [E]fforts to characterize descriptively, as well as to solve the problems of old age, at least in my opinion, should not take precedence over the pursuit of new knowledge regarding the mechanisms of aging [...]. Furthermore, it is a too often forgotten lesson of the history of science that the most significant societal advances usually result not by administrative design of the funding agencies, but instead, serendipitously from straightforward, non-targeted, investigator-initiated, high quality basic research.
>
> (Adelman, 1987: 227–229)

Adelman, having served as President of the Gerontological Society between 1986 and 1987, had become known for advocating, to funders and policy makers, a focused approach to the development of basic definitions of ageing, supported by fundamental research within and across diverse disciplines (Achenbaum, 1995: 130–131). In this, he was concerned that, since its establishment, the NIA had shifted its orientation from researching the mechanisms of ageing to focusing on the diseases of ageing, a process he later described as the 'Alzheimerization of aging' (Adelman, 1995; see also Chapter 9). His suggestion was that the biomarkers research programme partook in these troubling epistemic choices, where policy makers attempt to direct research towards societal and economic impact. These policy-driven research policies, he argued, were often ignorant of the role of serendipity in the 'history of science'. In this invocation of the 'history of science', and of the value of curiosity-driven research, Adelman also set forth a vision of division of labour within the biological and biomedical sciences, whereby basic research is institutionally separated from the development of technological innovation and the application of research in the clinic by qualified physicians. Blurring the laboratory and the clinic would lead, in his view, to unnecessary affective burdens on the activities of scientists and a resulting lack of focus on 'the pursuit of new knowledge regarding the mechanisms of aging'. The socio-economic and biomedical imperatives for developing a measure of BA did not thus justify the implementation of 'stop-gap measures' (Adelman in Haynes and Feinleib, 1980b) in ageing research, particularly if the pragmatic weaknesses of the BA approach, identified by Costa and McCrae, were to be considered.

The years following the end of the Biomarkers of Aging programme are acknowledged to have been difficult for BA proponents (Butler *et al.*, 2004; interview with Richard Miller, 23 January 2013), having lost support not only from previous political allies but also from within their own side. Responding to this, BA proponents shifted their attention from federal funding to gathering support and funding from independent, privately funded research institutes and foundations, such as the International Longevity Centre, the Ellison Medical Foundation or the American Foundation for Aging Research. Another important strategy appears to have been to make the critique of BA a point of departure for new research programmes. Indeed, in an interview I conducted with cell biologist Thomas von Zglinicki (20 June 2013), much emphasis was put on ensuring that possible BA measurements were robustly tested against the claims of critics before publicising any advancement in research in this field. In this, BA proponents' strategy appeared to revolve around honing the public presentation of their case.

Just two years after the failed NIA programme ended, Dr Richard Miller, a molecular biologist at the University of Michigan, argued that,

> The anti-biomarker camp is heartened by deconstructions (Costa and McCrae, 1980) of previous attempts to develop mathematical composites of age-sensitive traits that could be mistaken for quantifiable indices of 'generalized aging.' The critics of linear regression to biomarker construction argue, I think convincingly, that evidence for some correlation among different age-sensitive traits does not constitute proof that these traits are suitable indices for some unmeasured (and perhaps unmeasurable) 'rate of aging.' Previous attempts to develop batteries of biomarker assays have also stumbled over statistical obstacles, including failure to control for age-independent differences among subjects in the traits of interest. [But] in my view, though, these critics jump too quickly from 'has not been done' to 'cannot be done.' At the heart of their argument is the allegation that biomarkers of aging cannot be developed because 'there is no single rate of aging' […]. These critics may be right, but the question is too important to be decided by edict rather than by experimentation.
>
> (Miller, 2001: 21)

Acknowledging the methodological and statistical difficulties encountered by models of BA in the past, Miller repositions Costa and McCrae's critique as a 'deconstruction'. Denouncing the basis of the critique as the belief that 'there is no single rate of aging', Miller contrasts this 'allegation' with a scientific attitude of scepticism, openness to evidence and an experimental approach. This is directly linked to Miller's acknowledgement of the challenges faced by BA models and reinforces Miller's aim to portray the dispute as impersonal. In drawing on an established normative ideal of scientific conduct (Shapin, 2008), Miller was also aiming to condemn the irrational, interested ground upon which Costa and McCrae had based their defence of the institutional, administrative apparatus of CA, which could only be justified by 'edict' and the power of authority. In this denunciation, it was not insignificant that Costa and McCrae's original research had been partially funded by the Council for Tobacco Research USA, an organisation known for its pursuit of a defence of the status quo, in particular of the inelastic link between disability and mortality.

From this perspective, Miller's intervention can be seen as a direct response to the 'anti-biomarker camp's' defence of CA and its democratic, egalitarian nature. For BA proponents, such defence amounted to an undermining of the foundations of an open, pluralistic society, based on the pursuit of knowledge and experimentation. Following this line of normative justification for BA research, Miller and his supporters came to articulate a more consistent critique of what they saw as the political obstacles to their programme. This led them to a critical analysis of biomedicine, which they saw as eminently reliant on CA, and contrary to the research and societal agenda of health, activity and technology. However, instead of, like Adelman, condemning the incursion of ageing research into the biomedical model of R&D, BA proponents highlighted the failures of such a model in delivering the technoscientific promises of the ageing society (Miller, 2009). Against Costa and McCrae's proposal that the focus should be to understand 'for every case [...] the real etiology of age-related changes' proponents of BA pointed to the weak – 'health expectancy' related – returns from the biomedical approach of focusing on specific aetiologies, in fields like cancer or heart disease (see Chapter 9).

This line of attack on biomedicine's institutional apparatus was fruitful because it resonated against a backdrop of other condemnations of biomedicine for its decreasing rate of therapeutic innovation (e.g. Fanu, 1999). Importantly, it opened the ground for aligning new actors in a renewed approach to ageing, health and illness, such as the Medical Research Council, the Wellcome Trust (2006) or the European Commission (MARK AGE Program 2008–2013). As I will suggest in Chapter 9, this critique of biomedicine strategically positioned ageing researchers apart both from rejuvenation activists and 'futurologists' (Turner, 2009) and from the 'disease mongering' practices of the medico-industrial complex (Moynihan et al., 2002). Instead of promising to cure ageing or disease, BA proponents portrayed themselves as part of a scientific focus on ageing research that would bring about preventative, health-enhancing interventions and technologies (Butler et al., 2008).

But this attack on biomedicine is also a potential weakness, as the transformation of health can only be actualised if the elusive age measurement can ever be found. This is recognised by the 'anti-biomarker camp', and no significant attempts have been made to redress the balance in the controversy in the last decade or so (Robert McCrae, personal communication). For them, the 'onus is [still] on the other side'. Because no biomarker of ageing or battery of tests 'have so far proven [to be] a true measure of the underlying aging process', (AFAR, 2011) it is still the case that it is possible to deliver health and individualised care if the 'physician [is able] to assess the status of the kidneys, heart, and liver, and to evaluate them relative to age, sex, and other norms' (Costa and McRae,

1988: 213). For opponents of BA, standards or norms of health and function are there to assist physicians in their diagnostic and prognostic assessments, providing pastoral guidance for the patient in front of him/her. This can be done with and through the existing institutions of biomedicine.

Presently, the controversy appears to have reached a stalemate, with each 'camp' entrenched in their respective institutional configurations. On one side, there is a network of actors proposing to continue to use CA in combination with information on individuals' physiological parameters, everyday activities and 'attitudes' in order to clarify processes relating to the onset of age-related illnesses such as heart disease or stroke. Their focus is on establishing normative 'risk' parameters for specific age groups clustered to specific etiologies. This is the world of the new pastoral powers described by Rabinow and Rose (2006). On the other side of the debate, there is the ambition to establish new methodological conventions to be able to measure and chart biological age. Actors on this side of the controversy argue that such individualised ranking measurements should support an alternative configuration of welfare and health care institutions. Experts are distally positioned and knowledge mainly embodied in standards. Their vision is that this is the ground on which the technological promises of the ageing society can be delivered, by replacing the current reliance on professionals and the present research infrastructure with reliable measurements and a new focus on how the relationship between ageing and health is investigated.

Conclusion

Taking as a point of departure the problematisation of the relationship between ageing, illness and death that is concurrent with 'ageing society' assemblage, this chapter explored how age measurement became the focus of collective investigation. I have suggested that, in this process, widespread and trusting institutional reliance on CA to ascertain age-specific norms, values and expectations has been replaced by emerging epistemic and normative uncertainty about CA as a 'variable' and/or marker for social and political rights and duties. However, CA has not been substituted by another, singular, stable age standard but instead there is a proliferation of intended 'personalised' age standards developed and highly debated across a variety of forms of knowledge, decentralised organisations and purposes. These two processes have combined to produce torquing effects between CA and individuals' trajectories, and between alternative age measurements such as 'biological age'.

In relation to this, I have argued that uncertainties in biological age validation and implementation have drawn on and reinforced two modes of organising knowledge production to meet the challenges of the 'ageing society': one that relies on CA as a variable to construct niche clusters of health risk (Epstein, 2008), and on biomedical expertise as a guide to interpret data and guide individual conduct (Rabinow and Rose, 2006), and another which aims to embody in the alternative standard itself the mechanisms of individualisation of health, work and technology. These divergences concern not only CA or BA themselves but more importantly how to organise institutions to measure, monitor and manage age in contemporary societies, and help us understand, from an infrastructural point of view, the processes of deinstitutionalisation or individualisation of the life course that have been the concern of sociologists of ageing in the past decades.

Chapter 6

Individualising ageing?

Introduction

In the last chapter, I have explored how administrative and epistemic reliance on chronological age measurement has been increasingly challenged by approaches that emphasise the individualised nature of the ageing process. I have suggested that, in this regard, age measurement partakes in wider processes of transformation of the role of standards in social and economic processes in late modernity. In aiming to differentiate capacities, needs and desires, alternative age measurements have attempted to personalise the relationship between individual characteristics and work, or health and welfare services. In so doing, they have drawn on the figure of the individual as both an epistemic object and an actor – a subject of normative action. This is possible because ageing – it is generally agreed – is the process by which genetic and environmental elements, including lifestyle behaviours, combine in a highly individualised trajectory of biological and psychosocial change (e.g. Walker, 2014).

Within ageing research, this consensus is encapsulated, for example, in the concept of 'differential ageing' to express the range of probable individualised maturing pathways in any given historical context. This concept, and similar ones, express the idea that while there are common patterns of decline and adjustment associated with ageing – sometimes called 'normal ageing' – there is also 'enormous individual variation [,] age changes [being] highly specific' (Hayflick, 1998: 141). This dynamic speaks directly to debates about the relationship between ageing and health that I have argued are at the core of the agencement/assemblage of the 'ageing society'. Indeed, one of the key propositions in that controversy, the compression of morbidity hypothesis, is specifically reliant on assuming an inherent plasticity relating to the individualised timing and extent of age-related illness and disability (Fries, 1980). In policy circles, the individualised character of the ageing process is equally paramount. Policies, it is argued, should cater for such individual variability rather than enforce age-graded rights and responsibilities. Recognising the individual as the centre of the policy process has thus come to underpin active and healthy ageing policies (e.g. Eurostat, 2012). How did this consensus come about? How did the individual come to be seen as the pivotal, principal entity in understanding and managing the relationship between ageing, health and illness?

Answering this question entails exploring the epistemic, methodological underpinnings of this relationship, and in particular the establishment of an 'elective affinity' between the longitudinal method and ageing processes. The longitudinal method – sometimes labelled cohort study – is a type of observational study design that relies on the serial measurement or monitoring of individuals at different points in time. Its origins lay with the consolidation of the child development movement of the 1920s and 1930s, discussed in the last chapter, and its search for new knowledge foundations – normative standards of 'normal development' – to ground what Frank once labelled 'new designs for living'. It came again into prominence around World War II, with the establishment of community studies such as the Framingham Heart Study, which are usually linked to the development of contemporary risk factor epidemiology (Rothstein, 2003; Parascandola, 2004; Amsterdamska, 2005; Berlivet, 2005; Oppenheimer, 2006; Giroux, 2008) and programmes of epidemiological surveillance and health maintenance (Armstrong, 1995). One decade later, a variety of longitudinal studies of ageing – such as Duke's or the National Institute of Mental Health Human Aging Study – came to focus on what 'investigators often call[ed] the "dynamics" of aging', that is to say on 'individual differences in the patterns of physiological decline or psychosocial change, and the dependence of such patterns on social context and past history' (Bookstein and Achenbaum, 1993: 29).

These different deployments of the longitudinal method – and of the entity of the individual within – were not historically phased, and became instead articulated as different sets of epistemic practices, devices and norms, particularly as understanding the relationship between ageing, health and life expectancy became more pressing. Nowhere is this tension more evident than in the evolution of the BLSA, a major programme of investigation into the nature of the ageing process which the US National Institute of Aging has funded from 1958 to the present day. Headed for 25 years by Nathan Shock, himself an active researcher in child development research in the 1930s, the BLSA, like other longitudinal studies of ageing, was shaped by the interplay between these different agencements of the ageing individual. On the one side, there were those who emphasised randomisation and sampling to evidence claims about, and justify policies with respect to, the aetiology of disease. These relied on age group averages and regressions to the mean to devise normative standards for policy and practice. They were Queteletian in their approach to ageing and illness, aiming to use population data to calculate a 'true value' (Hacking, 1990; Desrosières, 2008: 130–138). Here, individuals are seen to be formed by the combination of shared characteristics, along multiple variance curves. On the other side, most notably headed by Nathan Shock himself, there were those who evoked the technical repertoire of physiological research, especially the notion of the 'model organism' – a purified, simplified workable exemplar, normally contained within laboratory-like conditions. Their practices can thus be located in the articulation between 'biological age' and 'uniqueness' described in the previous chapter, in that they viewed unusual, extraordinary cases as epistemically valuable exactly because they lacked 'representativeness'. In this agencement, the individual who is brought to bear is a singular, unique person.

In the chapter, I first describe the myriad sociotechnical strands that enabled the establishment of the BLSA, to then examine how different enactments of the relationship between knowledge making procedures and the management of ageing-related processes interacted to shape the current consensus on the individual nature of that process. In this, I am arguing that this consensus was established because, and not despite, of the unstable and contested meaning of the 'individual' with longitudinal studies of ageing and health. This fluidity is of consequence to our understanding of the epistemic infrastructure of pervading individualisation processes that sociologists of the life course have otherwise linked to economic globalisation, labour market deregulation and the restructuration of public services since the 1980s.

Establishing the BLSA

The official history of the BLSA traces its origins to the 'chance meeting' between Nathan Shock and William Peters, a retired officer of the United States Public Health Service (USPHS) who deployed his social networks to recruit the first batch of participants (Shock et al., 1984: 45–46). While most probably factually correct, such a history does not acknowledge the longer process that locates the origins of the BLSA at the intersection of three interlinked processes. The first of these was the translocation of the search for standards of physiological age from childhood to old age during the 1930s, as analysed in the previous chapter. Second were the consequences of such a translocation for crucial debates about the nature and focus of gerontology. The third process resulted from the attempt to resolve these debates, by aligning gerontology with the emerging agenda of public health during the 1950s, such that the longitudinal study was to become a key mediator. I expand on these below.

As discussed in previous chapters, in the first decades of the twentieth century Progressive Era reforms encapsulated an increased interest in child development from both privately and publicly funded organisations, such as the Laura Spelman Rockefeller Memorial Fund (LSRMF) (Smuts, 2008). Focusing on the need to characterise the 'normal' child, a variety of studies were established to examine the phenomenon of development by means of serial observations of selected children, key examples of which would be the Harvard Growth Study or the Oakland Growth Study (Sontag, 1971). As I argued, the knowledge making and normative ideals of what came to be known as the 'child development research movement' were embodied in the figure of Lawrence Frank, who as Director of the LSRMF pursued a vision that the scientific understanding of 'unadulterated', pure processes of physiological and psychological development should underpin the designs of social institutions and the management of individual behaviour (Sealander, 1997: 79–99).

Towards the end of Frank's tenure at the LSRMF, the Institute of Child Welfare Research at the University of California hired a then young physiologist by the name of Nathan Shock. In this setting, over the next ten years, Shock's research was to focus on physiological changes in adolescence based on serial measurement of girls in the Oakland public school system (Shock, 1943; Prescott, 2004: 168–169). Having privileged access to this residentially stable population, Shock had been able to establish that the onset of a physiological event – menarche – was more important than 'chronological age' in structuring changes in development. Such a discovery fully sanctioned the premises of the child development research movement, and, not surprisingly, led to widespread recognition of Shock's ability as a physiologist by both his peers and policy makers like Frank.

However welcome, such recognition had been decelerated by the slow and dispiriting progress of longitudinal research, where able and promising researchers were becoming increasingly frustrated also by 'low status unrewarding jobs' (Sontag, 1971: 991). By the turn of the 1940s, the problematic status of child development research was further challenged by shifts in the international economic situation and US policy responses to it. Indeed, as Shock had joined the Institute of Child Welfare Research, American society was beginning to experience the consequences of the Great Depression, making growing unemployment a key policy concern. The ensuing public debates over the introduction of some form of national insurance – which eventually resulted in the Social Security Act of 1935 – drew attention to the position of the elderly in society. As described in Chapter 4, this was framed around the issue of the declining quality of populations and a reimagining of the 'worth' of older people in society.

In this context, Frank was again called to lead the institutional apparatus for knowledge making in later life, as he had done for children in the previous decades. As vice-president of the Josiah Macy Foundation, he shaped a nascent gerontological programme within the USPHS, transporting into the programme concepts and methods of child development research, particularly the idea that it was necessary to understand 'normal' physiological processes to be able to generate adequate 'designs for living' in later life (see Chapters 1 and 5; Achenbaum, 1995: 76–79). Not surprisingly, given his difficult situation at the Institute of Child Welfare Research, Shock was quick to join this emerging field of research and, in 1941, through Frank's suggestion, replaced Stieglitz as Director of a newly established USPHS Unit of Gerontology (Park, 2016).

Establishing a programme of research in the emerging field of gerontology in the years that followed was a challenging task, as the interdisciplinary programme on which gerontology was founded struggled to find common ground (Achenbaum, 1995). Recognising the central position he occupied at the Unit for Gerontology, Shock drew on his experience as a physiologist to attempt to build consensus around what he saw as a shared epistemic and political entity: that of the individual. He argued that it was not sufficient to look at ageing solely at the level of cells and tissues, or simply as a population problem, because 'if you are going to explain ageing as a process [...] ultimately you have to look at individuals [...] from different points of view and at different levels of organization' (Wolstenholme and Cameron, 1955: 244). Individuals represented an ideal epistemic solution to the controversies within gerontology by deploying a methodological articulation – based on a 'natural unit' – between 'points of view' and 'levels of organisation'.

Individuals were furthermore at the core of Shock's political vision for the 'problem of ageing', where generalist age-grade policies of retirement were ignoring wide physiologically-based variations of individual capacity and need (Shock, 1947). Such a vision had become increasingly relevant as the political debate about the establishment of a federally funded health care programme became focused on older people and social security payments during the late 1950s (Oberlander, 2003). In this context, when in 1956 President Dwight Eisenhower established the Federal Council on Aging, Shock was able to translate its individualist and voluntarist thrust (Lockett, 1983: 71–72) by appealing for a recognition of 'older persons as individuals, not as a class, and their wide differences in needs, desires and capacities' (NWS: 21: 'Progress in the field of gerontology', 13 July 1956; see also Chapter 3).

In other words, for Shock, both understanding the ageing process and designing institutions for the elderly required a focus on the individual. To capture this unit of analysis and policy intervention, Shock explicitly drew on his own experience in child development research to propose that 'we should ... collect observations on the same individual throughout his [sic] life span' (Shock, 1956: 269–270). He suggested that, as in child development research, serial measurement of individuals should enable understanding of the relative contributions of what he then labelled 'pure ageing' and external environmental factors in the ageing process. This would be deployed through inter-individual and intra-individual comparisons to disentangle factors in the 'dynamic of ageing'. Only this knowledge base would support experts in the creation of the 'designs for living' his mentor Frank imagined as the solution to the 'problems of ageing'. Shock had been trying to address this methodological problem since taking over at the Baltimore Gerontology Branch (Park, 2016) but fruitlessly: the data he had been collecting with the assistance of the Baltimore Department of Public Welfare was not adequate for this purpose, as his human subjects were recruited from an institutionalised population and thus were 'not free from clinical signs or history of diseases' (Davies and Shock, 1950: 496). Those subjects presented, for Shock and his colleagues, a biased picture of the ageing process which could not be translated into the envisioned public policy reforms.

Hence, when, in 1957, William Peters offered to recruit his neighbours in Scientists' Cliff – a wealthy retirement community on the shores of Chesapeake Bay – as research subjects, Shock's assistant, Arthur Norris, was quick to realise that this was a unique opportunity to balance the picture of the ageing process (NWS: 21: Norris to Shock, 6 November 1957). The political implications of this opportunity were later articulated by the Secretary of the Department of Health, Education, and Welfare, in saying that,

> Until recently (October 1957), the Gerontology Branch, in its clinical research, used individuals of different ages (20–90 years) as research subjects. These people were medically indigent, admitted either to the Baltimore City Hospitals or the Old People's Home. Thus they did not represent the great majority of the aged who live normal lives in their own communities. Dr. Shock thus felt the need for more balanced research – another side of the aging coin. This need gave birth to a unique program of research on the aging process, now in its second year, the Scientists' Cliff Project.
>
> (NWS: 22: 'The aging program of the National Institute of Health', July 1959)

While Shock was initially ambivalent about the Secretary's naming of the project after a particular location (NWS: 22: Shock's annotation of 'The aging program of the National Institute of Health', July 1959), he could not have objected to the political recognition that there was 'another side of the aging coin' which could only be reached by studying individuals 'who live normal lives in their communities'. Reforms of health and welfare, as Shock himself put it two years later, were to come 'from studies in which the laboratory is, in effect, the community' (Shock, 1961: 102). The BLSA was thus born as the exemplar of a research programme which by focusing on community-dwelling 'individuals' promised to solve the problems presented by an ageing population.

Twenty-five years later, Shock and his colleagues at the BLSA were still devoted to the use of the 'community laboratory' to distinguish,

> the true effects of aging and those processes, including disease, socioeconomic disadvantage, and lack of educational opportunity, that may also appear or become more pronounced with time but are biologically irrelevant to the underlying mechanisms of human aging.
>
> (Shock *et al.*, 1984: 1)

How the BLSA could deliver such a result after 25 years was not a straightforward matter because while,

> [a] primary concern in the selection of subjects for a longitudinal study is their commitment to continued participation in the study and their geographic stability over long periods, [t]his requirement limits the sampling procedures that can be used and must be taken into consideration in the generalization of conclusions drawn from any longitudinal study. [...] Subjects selected at random will show both a high initial rate of refusal to participate

and a high drop-out rate when presented with a test schedule that includes uncomfortable and time-consuming procedures. To this extent most longitudinal studies, including the BLSA, have compromised true representativeness in order to obtain loyalty and cooperation from their subjects.

(Shock *et al.*, 1984: 14)

This is a curious statement because it seems to be aimed at undermining the value of longitudinal studies in understanding ageing or health, as their results cannot be easily generalised to the population. From this perspective, such a statement would appear to be a recognition of the failure of the longitudinal study to deliver the promises Shock had embodied in it 25 years before. However, a closer look at the quote reveals instead that Shock was arguing for a specific understanding of studies of individuals in 'which the laboratory is, in effect, the community'. His argument was that the value of representativeness was somehow in tension exactly with what made longitudinal studies unique and epistemically valuable: their ability to depict individual changes across time points. This uniqueness relied, Shock suggested, on 'loyalty and cooperation' from research participants, a form of commitment that enabled understanding individuals living in their community as laboratories. It was exactly this configuration of knowledge making that resulted in caution having to be taken 'in generalisation of conclusions from any longitudinal study'. Willingness or capacity to participate in longitudinal studies was not evenly distributed across the population, resulting in samples that were skewed and not representative of the 'average man'.

Shock's statement was thus an attempt to position longitudinal studies within the wider landscape of populations-based studies, which derived much of their value from the capacity of sampling procedures to enable calculations of distributions, averages, etc. As Marks (1997) has suggested, in the US this process was driven by a set of reforming biometricians within the National Institutes of Health (NIH), who were particularly active in establishing the methodological standards for federally funded population research, overseeing the design of studies, or, in cases such as the BLSA, arguing for its improvement and reorganisation of sampling and recruitment. Such focus on knowledge procedures was reinforced by the increased requirement to obtain adequate representation of different groups within the population (Epstein, 2008), such that by 1984 representativeness conflated social, political and symbolic meanings in public calls to reform medical research.

Against such a combination of methodological and political considerations, Shock and his colleagues offered the argument that the value of longitudinal studies was dependent upon finding a group of participants that were not only willing to sustain a longstanding relationship with a study but also possessed other data-related qualities:

[participants'] above average educational and socio-economic status [...] offered a number of advantages for the study of aging: improved accuracy of such information as family history, past history of illness (etc.), accuracy of activity history, accuracy of nutritional diaries [...] and ability to understand and execute a variety of questionnaires, and increased probability that the subject would be interested in the scientific questions addressed by the study. In short, this selection of subjects provided the opportunity to study the effects of aging under conditions in which economic status and educational level would not serious limit medical care, nutrition and other factors affecting health over the life span.

(Shock *et al.*, 1984: 52)

In other words, the BLSA investigators were arguing that it was exactly the special, non-representative characteristics of the subjects recruited into the BLSA which made them ideal for the study of ageing. These subjects' privileged social and educational status not only fitted the methodological requirements of the longitudinal approach such as geographical stability and the ability to provide complex information, but also served to exclude processes which Shock and his colleagues regarded as essentially 'biologically irrelevant to the underlying mechanisms of human aging'.

Shock's and his colleagues' defence of the BLSA methodological approach is striking in light of what we know about the importance of representative samples for generating valid, robust and relevant population research. Challenging this epistemic and political consensus, they were effectively claiming that knowledge produced by longitudinal studies could not be warranted by the dominant epistemological and methodological standards of population-based research, but rested instead on an alternative set of norms for the production of knowledge of ageing and health. This is all the more astonishing as the BLSA investigators were claiming an exceptional status for a major federally funded programme of research, in a context where epidemiology was beginning to shape public policy and clinical practice, and where recruitment into epidemiological studies was regulated ever more tightly by methodological and political standards, as I suggested above.

To understand how the BLSA could claim to be so immune to the dominant epistemic norms, as well as regulations and policies governing biomedical population-based research, it is necessary to explore the trajectory of the BLSA within the epistemic and institutional contexts of epidemiology and ageing research from the 1960s to the 1980s. In these years, we will be able to find the origins of its methodological and political 'exceptionalism', that is to say, of the BLSA's deviant position in relation to standard approaches to understanding ageing in populations. In analysing four episodes in the years between the formation of BLSA and the 1980s, it becomes clear how this exceptionalism is intimately linked to how different articulations of the individual as an epistemic and political entity coexisted within the BLSA and longitudinal studies of ageing more generally.

Episode 1: Colloquium on Longitudinal Studies, 1965

In the early 1960s, the scientific maturing of research on development and the increased political significance of ageing in light of discussions around what was to become Medicare combined to relocate the various subsections of the specialist, constitutive institutes of the NIH that were dedicated to either children or the elderly into a new structure: the National Institute of Child Health and Human Development (NICHD) (see Lockett, 1983: 59–63). This organisation was specifically created to focus 'on the complex health problems and requirements of the whole person rather than on any one disease or part of the body' (NWS: 23: Invitation to attend Colloquium on Longitudinal Studies, 2 December 1964, my emphasis). Given the historical antecedents of research into non-child development and ageing, described above, the longitudinal approach was quickly identified as one of the key areas of attention, leading to the organisation of a conference on the topic.

The organisers of the NICHD Colloquium on Longitudinal Studies, including Shock himself, were concerned that many longitudinal studies had been set up solely as the result of individual researchers' particular interests. This resulted in a weakness of the research field, such that participants agreed that '[the] principal handicap in longitudinal research [was] the lack of integration, uniformity and relatedness among the work of researchers concerned with various age groups' (NWS: 23: Invitation to attend Colloquium on Longitudinal Studies, 2 December 1964). There was also, however, uncertainty about the methodological challenges presented by the integration of such studies into one common, standardised approach.

Diagnosing this problem, Samuel Greenhouse, then a statistician at the NIH National Institute of Mental Health (NIMH) – and one of the key methodological reformers of population research in the US (Marks, 1997: 157) – observed that, 'the analysis of longitudinal data has been a makeshift hodgepodge', adding that,

> what [most researchers in the field of longitudinal research] seem to be doing ... is using common sense and non-probabilistic, non-mathematically rounded procedures[;] in the absence of anything better they are using conventional methods, and usually become aware of it rather quickly when their conclusions are wrong.
>
> (NWS: 23: Colloquium on Longitudinal Studies: 141)

In this statement, Greenhouse was overtly critical about the methodological procedures followed by most if not all researchers in the field of longitudinal studies of development and ageing. He characterised them as relying on common sense conventions – rather than probabilistic reasoning or mathematics – with the result that most findings and conclusions drawn from longitudinal studies were necessarily wrong. His proposal was that it was necessary to improve the methodological underpinnings of longitudinal research, particularly through the use of probabilistic procedures to sample populations. This was required not only for scientific reasons but also because, if the results of such research were to be of any use for general public policy, it was necessary to make them mirror the underlying, 'true' structure of the population it was addressing. Longitudinal studies, Greenhouse appeared to be saying, stood outside the central paradigms of population science, as described in Chapters 3 and 4.

This assessment was not to be taken lightly. As one of the first recruits into the Division of Statistical Methods at USPHS, Greenhouse had been involved in the development and refinement of statistical methods in biomedical and population research for almost two decades. His work on how to address divergences from assumptions in repeated measures using analysis of variance – the Greenhouse-Geisser correction – was particularly renowned among professional and academic statisticians. Further, as co-author of *Human Aging*, he could speak with some authority about the study of physiological and psychological changes in the ageing individual (see Birren *et al.*, 1963). Thus, he was not alone in this assessment of the deficiencies of existing longitudinal studies such as the BLSA. Sam Shapiro, a statistician then working for the Health Insurance Plan of Greater New York, added to Greenhouse's critical assessment of longitudinal research, suggesting that it appeared that in most cases 'a particular group of subjects … [was] often chosen more because of availability than appropriateness for the purpose of the study or the relationship to broader population groups or fundamental properties of man' (NWS: 23: Colloquium on Longitudinal Studies, p. 134). In other words, convenience was trumping representativeness.

These critical statements, however authoritative, did not go unchallenged. Some participants in the colloquium, particularly those directly involved in conducting longitudinal studies, suggested that 'random samples are not necessary when we are concerned with establishing [physiological] relationships; they are more important when we are making estimates' (NWS: 23: Colloquium on Longitudinal Studies, p. 136). From this perspective, there were key differences in knowledge making standards between those applied to investigations aiming to understand the basic mechanisms of ageing and those focused on gauging the strength and significance of such relationships in shaping the ageing of populations. Their argument was that while randomisation and analysis of error might be important when trying to understand population phenomena, when focusing on physiological relationships – or what Shapiro called the 'fundamental properties of man' – other methodological criteria applied. As one of the participants in the colloquium strikingly put it,

> *Our sample isn't representative … that's why we selected it.* We are not trying to present averages or means or patterns of the entire population of the world…. We have a laboratory population in which we are trying to study certain mechanisms of growth, the development of patterns of behaviour, etc.,… It depends on what you are trying to study…. We are not studying areas in which a representative sample is of any consequence…. What you want is a homogeneous population.
>
> (NWS: 23: Colloquium on Longitudinal Studies, p. 136; emphasis added)

This statement encapsulates a whole challenge to the methodological standardisation proposed by Greenhouse, Shapiro and their network of statisticians. According to this unidentified speaker, the selection of populations for longitudinal studies should be dictated by substantive criteria rather than by methodological standards. This is because there should be a fit between your population selection procedures and 'what you are trying to study'. Physiologists were concerned with establishing 'certain mechanisms of growth [or] the development of patterns of behaviour'. To do this they required a 'homogeneous' group' – a 'laboratory population'. This assertion positioned Greenhouse and other statisticians in an auxiliary and subordinate position in relation to physiological and psychological researchers interested in basic mechanisms of ageing. In establishing these relationships and the best population to study it, statisticians were best advised to assist, rather than impose their own epistemic considerations. Thus, it was argued, in studies aiming to understand patterns in diverse populations and in making estimates, randomisation, representativeness and the expertise of statisticians might be important, but 'to study certain mechanisms of growth' what was required was 'a laboratory population'.

Researchers associated with the BLSA were acutely aware of this tension between aiming to understand fundamental biological mechanisms and methodological prescriptions, between physiological models and statistical averages and differences. As early as 1959, Shock himself had admitted that 'ideally, we should study a random or stratified sample of the entire population' (NWS: 22: 'Physiological considerations in the care of the aged', 27 May 1959). However, in establishing the BLSA, he and his colleagues were aware that the group of selected participants could not, as a whole, meet that ideal. In effect, they had been recruited to balance the representations of ageing obtained from populations of institutionalised subjects such as those linked to the Baltimore Hospitals (see above). As Stone and Norris (1966) observed in a later review of the subjects recruited into the study, the BLSA community of participants represented a homogeneous population who were distinctive in their social and economic make-up. As a group of 'blue ribbon breadwinners' selected by 'word of mouth', as the *Washington Post* would put it (NWS: 22: 'Aging study pair get 3 days of probing', Washington Post, 30 August 1959), the population recruited into the BLSA could be described more accurately as a 'laboratory population', a homogeneous population uniquely suited to meet the requirements of physiological research on ageing.

Episode 2: Review of the Longitudinal Study of Aging, 1971

In 1971, as the implications of the establishment of Medicare programmes were being publicly articulated (Oberlander, 2003), the organisation of research on ageing, including the investments of the NIH and the position of the renamed and greatly expanded Program on Aging of the NICHD, again became a matter of political debate within the United States Senate (Lockett, 1983: 95–102). As representatives aimed to find a new institutional framing of gerontology that was congruent with the new health care programmes, a re-evaluation of the federal research on ageing came quickly on the cards, and a review of the BLSA was subsequently scheduled. This review was predictably to reopen the questions and tensions that had shaped the Colloquium on Longitudinal Studies, just six years before.

In particular, the review commissioners had specifically emphasised that special attention should be paid 'to the structure of the cohort of the longitudinal study' (NWS: 24: Report of the Special Advisory Group, 14–15 June 1971). One especially problematic aspect of the population recruited into the BLSA was the absence of women. This concern was articulated by James Birren, chair of the review panel and former Director of the Program on Aging at NICHD, for whom such absence seemed to function as the marker for a more general unrepresentativeness of the BLSA. Birren, a former collaborator of Greenhouse, and by then one of the central figures in the psychology of ageing, embodied the view that only representative, diverse samples could provide researchers with adequate representations of the differential ageing phenomena (Birren, 1959).

Therein lay his divergence with Shock and his colleagues, who argued instead that the differential, individualised nature of ageing was best demonstrated in a homogenous group, disconnecting such variations from life chances and living conditions. Thus, in relation to the inclusion of women, for Shock, the issue remained one of substantive, physiological criteria, questioning 'what would the study of women add to scientific goals' of the study of the ageing process. From his perspective, accepting the 'sociological inclusion criteria' implied hypothesising that differences in ageing derive from social conditions rather than individual, physiological characteristics. In line with this position, Shock was only able to accept the addition of women to the sample when Edward Bierman, then Head of the Division of Metabolism at the University of Washington, suggested that there might be significant physiological differences between ageing in men and women – chiefly a different 'rate of bone loss' (NWS: 24: Study Review Committee: Notes, 14 June 1971).

The disagreement between Shock and the panel, most particularly Birren and Greenhouse, was not just about the hierarchical ordering of physiological and social variables, but also about the use and purpose of statistical methods. This is observable in Shock's correspondence relating to the composition of the panel, and in particular his acute concern about Warner Schaie's nomination and eventual appointment to the panel (NWS: 24: Notes, 10 December 1970). Schaie, then based in the Department of Psychology at West Virginia University, and most notably the founder of the Seattle Longitudinal Study, was, at the time, deeply involved in discussing and setting the methodological standards that should determine the organisation of longitudinal studies, standards in which representativeness and control groups played a pivotal role (e.g. Schaie, 1973; also Schaie and Baltes, 1975). He was particularly enthusiastic about the role of properly designed longitudinal studies in addressing the social problems of the elderly, in an approach that Shock himself would have seen as redolent of treating older people 'as a class' and not individuals (see above).

Shock, it turned out, was correct in his assessment of the bias of the panel towards methodological standards and the ideal of representativeness. He was not surprised therefore that when the panel recommended that 'consideration should be given to establishing a more representative sample of male and female cohorts as a scientific resource for NICHD', it also observed that the population recruited into the BLSA,

> is a selected segment of a markedly advantaged group, apparently quite homogeneous with reference to education and socio-economic status. At the upper age levels, it is markedly advantaged with reference to health, and is presumed to be markedly health conscious. For the study of physiological processes and relationships, the atypicality of the study sample may not be a serious problem. It is likely to be a problem, however, where social and psychological variables are involved.
>
> (NWS: 24: Report of the Special Advisory Group, 14–15 June 1971)

In this assessment, the panel appeared to revisit the divisions between physiological models and statistical samples already marked in 1965, only to rearticulate them. First, it conceded that the unique, homogeneous character of the population recruited 'may not be a serious problem' for the study of physiological processes, a point that Shock and his colleagues had been making since the foundations of the study. The uncertainty introduced into the proposition by the auxiliary 'may' is of some importance, however, because it contrasts with the problems that were 'likely' to be created by this approach for the analysis of social and psychological variables. In other words, the panel seemed to be suggesting that any positive physiological findings that might result from the study of such an unrepresentative sample could not be corrected in the study of the psychosocial understanding of the ageing process. These two divergent epistemic goals could not be reconciled due to their distinctive methodological requirements. Samples and models, the panel appeared to be suggesting, could not coexist in the BLSA.

This line of argument seemed also to intimate that the BLSA was out of line with the multidisciplinary ideal that gerontologists had attempted to construct, or at least, had publicly cultivated during the previous three decades (Achenbaum, 1995; Park, 2016). Shock, remember, had just a few years earlier proposed that the lack of integration in gerontology could be solved by implementing a focus on the individual as the key epistemic and political entity (see above). But the 1971 review panel doubted the grounds on which this vision could be built, as they were effectively suggesting that were the BLSA to be taken as a politically significant study of human ageing, it would have to move towards representativeness because only this methodological standard could secure access to both biological *and* psychosocial aspects of ageing. If gerontology was to have any impact on 'designs for living' it would have to be drawn on such a composite version of the ageing process. It would have to study individuals in their constituent dimensions of variables in a statistical space, rather than as unique entities.

This alignment of statistical, physiological and psychosocial considerations was persuasive, resulting in Shock agreeing to 'have closely attached to the project, on a permanent basis, an epidemiologist, who would have some expertise in demography and biometrics' (NWS: 24: Report of the Special Advisory Group, 14–15 June 1971). This was done in two stages. In the first instance, James Schlesselman, a statistician in the Biometry Branch at NICHD linked to Greenhouse, was recruited to provide advice on how to modify the sample so that statistically significant, intra-individual differences could be measured. Later, Clyde Martin, a trained social statistician, was appointed to manage Schlesselman's recommendations and to accompany the recruitment of new subjects (NWS: 24: Report of the Special Advisory Group, 7–8 December 1972).

Strikingly, however, the process of methodological reformation of the BSLA actually worked to reinforce its special character, its focus on a homogeneous 'laboratory population' and its idealisation of the individual. As Martin and Arthur Norris explained a few years later:

> In spite of the fortuitous manner of its recruitment, the resulting sample appears to be well suited to many kinds of investigations because of its high degree of homogeneity for such characteristics as socio-economic-status, health, colour and marital [...]. Moreover, since the focus of the investigation by the research staff concerns the search for interrelations between physiological, biochemical, psychological, pathological and behavioural variables, and how they may be related to age, the net result is to allow the social characteristics of participants to be disregarded as important sources of variation in most analyses.
>
> (NWS: 24: 'The BLSA: Characteristics of the sample', November 1976)

The fit between the population, the research focus and the statistical techniques used in the BLSA may have been fortuitous, but it now was robust enough to justify the continuation of the study and of its sampling procedures from a statistical point of view. The substantive focus of the BLSA on interrelations between 'physiological, biochemical, psychological, pathological and behavioural variables', on the one hand, and ageing, on the other, required special attention to intra-individual differences. From this perspective, the uniqueness and homogeneity of the population recruited into the BLSA was no longer a problem. This was because this uniqueness had been transformed into a method of 'variable control'. As a result, the BLSA sample was no longer statistically unjustified, and a threat to the epistemic unity of gerontology. In other words, the BLSA possessed the methodological means by which answers to its distinctive questions – questions focusing on the individual alone – could be integrated back into a multidisciplinary understanding of human ageing.

Episode 3: Review of the Baltimore Longitudinal Study of Aging, 1978

In 1974, the mounting political struggles over better care for older people resulted in the establishment of the NIA. Two years later, after intense political negotiations, Robert Butler became the first Director of the NIA and, at a time of growing calls to reduce federal expenditure, was immediately put under pressure to demonstrate the efficacy of federally funded gerontological research (Inglehart, 1978). Butler thus undertook a series of internal reviews of the research under his supervision, and the focus fell once again on the BLSA and its methodological design.

In this review, Butler brought to bear once more the alliances which he had forged during the evolution of the NIMH Human Aging Study and eventual production of *Human Aging* (Birren *et al.*, 1963), including an old critic of the BLSA, Samuel Greenhouse (see above). Greenhouse was by 1978 a much more established authority on methodological questions in population research, a reputation buttressed by support from the Director of the NIA (Butler, 2003). His aim was to question the very significance of the longitudinal method and its place in the investigation of ageing. This, however, was not a simple repetition of old qualms and objections, but a renewed challenge to the core claims of longitudinal studies as being able to capture the 'dynamics of ageing'. It was not only that longitudinal studies were expensive and difficult to manage, but also that results were slow to materialise and did not provide clear answers to the key question of the relationship between ageing, mortality and health – the central question of the 'ageing society'.

It was also the case that, with the replacement of Shock as the Director of the BLSA by Reuben Andres, such questions were less charged with professional criticism, as Shock's career was seen as representing the promise of the longitudinal study in the study of ageing (see above). This was explicitly noted by the panel, writing how Andres had,

> repeatedly impressed the Board with his detailed grasp of the multiple facets of the Longitudinal Study, with his obvious awareness of, and sensitivity to, all the criticisms that could be levied against the limited (elitist) sample or choice of variables for study.
> (NWS: 24: Baltimore Longitudinal Study of Aging, 18–19 May 1978)

This recognition was part payment for the efforts made at the BLSA to integrate statistical expertise, detailed above. It cast Andres' engagement with these issues as somewhat different from Shock's management, in his 'sensitivity' to the criticism directed at the methodological approach taken by the BLSA. But something else had changed. Whereas during Shock's tenure the scientific reputation of the study was guaranteed by its director's standing in the field of ageing research, reviewers were now concerned that data analyses produced by the BLSA had escaped proper scientific peer review:

The Board noted that the rate of publication of new data has accelerated over the last two years. Those in charge of the program are aware of the importance of publication in appropriate specialty journals rather than, or in addition to, journals devoted solely to gerontology, which escape criticism at the level of sophistication common in other fields, or books on aging, which may escape critical review altogether.

(NWS: 24: Baltimore Longitudinal Study of Aging, 18–19 May 1978)

The problem appeared to be that knowledge produced by the BLSA was not properly warranted. Publication in 'journals devoted solely to gerontology' was, according to the panel, not an adequate way to guarantee the quality of research. In effect, the reviewers appeared to go as far as to suggest that ageing research could no longer be contained in the discipline of gerontology, as it escaped 'the level of sophistication common in other fields'. This is understandable, as the delineation of ageing as an issue progressively shifted from being contained within a population group – the elderly – towards affecting the whole population. Given the increased relevance of ageing and the 'ageing society', the reviewers proposed that gerontology's approach was not a secure foundation to address such a wide issue. For the reviewers, most of them working outside gerontology, the quality of the research was problematic, it being uncertain whether 'several of the [projects] that were presented would, if they were to be reviewed by the peer review mechanism used for judging extramural research, be given a high enough priority to be funded' (NWS: 24: Baltimore Longitudinal Study of Aging, 18–19 May 1978).

One of the key problems in gerontology's epistemic weakness was its reliance on the hope that the longitudinal approach would provide access to a better understanding of the 'dynamics of ageing': the way in which individuals differentially experience decline and adjustment throughout the life course. This was not only a challenge to the BLSA but to gerontology's consensus that intra-individual variations across the time period would provide the key to understanding ageing. Reviewers instead wondered if 'similar or identical information' could be obtained by means of 'cross-sectional studies', 'short-term longitudinal studies', or 'experimental work' (NWS: 24: Baltimore Longitudinal Study of Aging, 18–19 May 1978). Even those among the panel who 'defended the strategy of the longitudinal study, strongly' suggested that perhaps the greatest contribution of the BLSA was to have demonstrated that longitudinal analysis 'gives the same results as cross-sectional analysis' (NWS: 24: Baltimore Longitudinal Study of Aging, 18–19 May 1978). Such an assessment amounted to a condemnation, as it suggested that the BLSA, once taken as the paradigm of longitudinal studies of ageing, exactly demonstrated the methodological worthlessness of longitudinal studies in general.

The panel's overt critique of the longitudinal study also called into question Shock and others' promise that the serial observation of individuals living in the community would provide the means to rebuild health and welfare institutions for older people. It questioned, most importantly, the authority of Shock's proposal that studying individuals' differential ageing was the best way to know and manage them as individuals. Instead, they proposed that cross-sectional studies or experimental work should provide the data necessary to integrate individuals into work, health and welfare programmes. But, here, individuals were imagined in a fundamentally different way: they represented unique configurations of variables, or points within multivariable distributions. In this regard, the panel's position is redolent of Costa and McCrae's (1980) critique of the concept of biological age, explored in the last chapter. Like Costa and McCrae, the panel argued that the methodological standards for the validation of populations-based research – statistical significance, error calculations, etc. – rendered the promise of a model of differential, individualised ageing unworkable. Furthermore, the tendency for contemporary social institutions to use standards to differentiate rather than homogenise would favour the aim to distinguish between levels or type of risk within populations, clustering the needs/abilities to plan for market formatting or service provision (Busch, 2011).

But all was not lost. If the BLSA demonstrated that the statistical comparison of different age groups yielded the same information as repeated measurement of the same individuals, this meant that the data the BLSA had generated was not in fact meaningless, but had been perhaps obtained at an unreasonable cost. This interpretation of the meaning of the BLSA data was however imbued with epistemic and political uncertainty, not least because of how it would impact on other longitudinal studies. Crucial in this uncertainty was the persuasive power of the BLSA approach to studying individuals as models of 'pure ageing', which relied on their choice of an extraordinary, non-representative 'laboratory population'. In this, BLSA participants had to be deployed as 'model organisms', that is to say, 'integrative models for whole, intact organisms [studied] using a variety of disciplinary approaches' (Ankeny and Leonelli, 2011: 316; see also Ankeny, 2008).

Because of this uncertainty, the 1978 review panel was again only able to propose another balance between methodological positions, and propose that 'it should be possible within the next 5–6 years to make a hard decision [about whether or not] to eliminate certain types of investigation from [the BLSA] or to phase out the [BLSA] completely' (NWS: 24: Baltimore Longitudinal Study of Aging, 18–19 May 1978). However, the panel had, for both micropolitical and methodological reasons, been able to initiate a critique of the longitudinal approach, and in particular its promise of generating knowledge of individual ageing for suitably individualised programmes. This signalled a change in the balance between versions of the individual at the BLSA which was to establish itself more securely from the early 1980s.

Episode 4: Reassessment of the Baltimore Longitudinal Study of Aging, 1987

In the early 1980s, following a series of reviews to intramural projects such as the BLSA, the NIA focus progressively emphasised the funding and support of extramural programmes, and of disease-focused research, most notably on Alzheimer's disease (Moreira, 2009; Chapter 9). These shifts meant that intramural research became increasingly subject to data-sharing demands across projects and investigators, with the consequent formalisation of the relationships between collaborating researchers, on the one hand, and between researchers and their human subjects on the other. They also signalled a change in the institutional position of the NIA, which was not only receiving public support for its research on Alzheimer's disease (Fox, 1989) but also for its achievements in the longitudinal study of ageing (Butler, 1983). Part of this acknowledgement was the public commemoration of the 25th anniversary of the BLSA, which had been endorsed by President Ronald Reagan (NWS: 24: Reagan to Shock, 7 September 1983) and eventually resulted in the publication of and international praise for *Normal Human Aging* (Shock *et al.*, 1984).

By 1987, this combination of factors created a predicament for the long-scheduled but, perhaps inevitably, belated review of the BLSA. If, on the one hand, the publication of *Normal Human Aging* had been mobilised by officials at the NIA as the ideal moment for the anticipated 'hard decision' about the future of the BLSA, discussed in the previous section, on the other hand, the anniversary of the BLSA raised public and political awareness of the importance of longitudinal data 'to help extend the healthy, productive middle years and to improve the quality of life for older Americans' (NWS: 24: The NIA's Baltimore Longitudinal Study of Aging, 1958–1983). Linking the BLSA explicitly to the problematic of the 'ageing society' and its agencement of health and morbidity reversed somewhat the momentum of critique that had been established just a decade earlier against longitudinal studies of ageing. But this capturing of the BLSA's aims was not without its problems, because the study had explicitly wanted to exclude questions of illness, in order to understand what Shock then labelled 'the other side of the ageing coin'.

The solution to this impasse was in the hands of Richard Greulich, who had replaced Nathan Shock as Director of the NIA Intramural Research Program. Reading the movement of the NIA towards disease-specific programmes as a shift in the orientation of ageing research, Greulich's proposal was to align the BLSA with the NIA's new focus on the understanding of the aetiology of age-related illnesses, and to progressively eradicate the aim to isolate 'normal' or 'pure ageing'. This represented a fundamental rethinking of the BLSA, which required extensive negotiation within the BLSA research team. As Greulich explained in a brief memorandum to the staff at the BLSA:

> During the course of writing Normal Human Aging, my colleagues and I developed a comprehensive statement of the BLSA objectives which were set forth in that volume [...]. It became increasingly clear that this statement needed to be changed, based both on our perceptions and from external critiques, most recently those of the [NIA] Board of Scientific Counsellors.

The earlier statement was deficient both in relying in part upon outmoded concepts and in creating unrealistic expectations about what the BLSA can and should achieve.

(NWS: 24: Revised BLSA Objectives, 4 February 1987)

Greulich's reference to 'outmoded concepts' and 'unrealistic expectations' is highly significant. As Shock's written annotations to this memorandum suggest, the NIA now seemed to regard gerontology's founding ambition to identify 'pure ageing' as no longer sustainable. In the new context of the 'ageing society', pure ageing was an outmoded concept, as it no longer addressed the more continuous, ambiguous relation between ageing, health and illness. The focus was now on the 'side of the aging coin' that Shock had wanted to avoid three decades earlier – age-associated illnesses and diseases. In such a context, disease-specific programmes now sustained the 'realistic' promise of illness-free and independent life in old age, and longitudinal studies of ageing could only, at best, be an auxiliary enterprise in this assemblage (see Chapter 9).

As referred to above, these new scientific objectives were accompanied by the consolidation of a new organisational framework whereby data collected in the BLSA could be accessed or requested by researchers across federally funded projects. In this new configuration, the BLSA would be best described as a data platform, in that it served as a data infrastructure for an extended, collective group of investigators, who were only aligned in their – centrally monitored – commitment to the central question of the 'ageing society'. For BLSA researchers, such an arrangement threatened the relationship between the researchers and the subjects recruited into the study. As Shock (1984) explained in his Introduction to *Normal Human Aging* (see above), the shift in the selection of participants to create a representative sample – with appropriate variations of health and illness – undermined what, after Rabinow (2005), could be labelled the 'biosocial' foundations of the BLSA, that is to say, the fact that the participants constituted a community rooted in shared somatic experiences of ageing. Thus, as early as 1983, Shock was questioning the changing relationship between the BLSA, its population and the intramural and extramural programmes in the following terms:

The … issue is whether the goals of the Baltimore Longitudinal Study of Aging remain as originally stated…. I must confess that the suggested changes in testing interval ('on demand' rather than a fixed interval and variable length of stay and selection of individuals to be tested by investigators) seem to me to be designed more for the convenience of investigators and the support of cross-sectional studies than to strengthen longitudinal studies…. If the longitudinal subjects are to become an 'animal colony' for the [Gerontology Research Centre], I think we should be honest with them and renegotiate our commitment to them and their commitment to us.

(NWS: 24: Shock to Greulich, 26 October 1983)

From Shock's perspective, the effective transformation of the BLSA into a 'panel platform' meant that the testing of subjects would vary according to the requirements of investigators not directly associated with the BLSA. Such variation not only undermined the longitudinal approach in its dependence on consistent and uniformly repeated testing, but each new requirement would also entail renegotiation of the relationship with the participants by means of 'informed consent' processes. This, Shock maintained, would transform the participants into an 'animal colony'. By mobilising this analogy, Shock was seeking to draw a boundary between the BLSA and the kind of experimental research favoured by the NIA. Subjects in longitudinal studies, and those involved in the BLSA in particular, Shock argued, were committed to the study's aims in a way that no animal colony could be. In fact, recruitment to the BLSA had required sharing the scientific aims with the participants, so that changing aims could not be a question for investigators alone, but had to involve the participants.

Shock's argument appears to have been persuasive because it was eventually agreed that 'initiation of a BLSA protocol by an outside investigator [should be conditional] on the existence of a collaborative relationship with an NIA intramural scientist' (NWS: 24: Baltimore Longitudinal Study of Aging Discussion Group: 28 March 1986). Securing the role of an intermediary between the NIA and 'outside investigators', on the one hand, and the population recruited into BLSA, on the other, was crucial to Shock, Andres and their colleagues because the claim to exclusive knowledge of how best to run a longitudinal study enabled them to question the ambitions of the NIA. Securing the role of gate-keepers enabled BLSA investigators to continue to negotiate the scientific aims of the study, and in particular how the individual as an entity was deployed within the guiding concepts of the study. Specifically, as Greulich sought to anchor the study of disease in an ageing population on the concept of 'successful ageing' (Rowe and Kahn, 1987), BLSA investigators objected that there were irreconcilable divergences between the design of the BLSA and an investigation of the prevention of disease and disability in later life.

During a meeting of the BLSA Discussion Group just a few days after the circulation of Greulich's memorandum mentioned above, Andres raised the stakes by noting how this focus on 'successful ageing' raised difficult questions about future patterns of recruitment into the BLSA, because the goals entailed two 'incompatible study designs' – 'a pure longitudinal study of aging processes' and a 'study of factors influencing longevity or mortality' (NWS: 24: Baltimore Longitudinal Study of Aging Discussion Group, 17 February 1987). Understanding the age-related determinants of disease and disability, Andres claimed, relied on the possibility of drawing statistically rigorous and reliable inferences from survivors into the eighth and ninth decades of life. This, in turn, required a 'rectangular' pattern of recruitment such that equally sized cohorts were maintained and observed for each of decade of life. As Andres noted, not only had such an ambitious ideal never been achieved by any of the many longitudinal studies of ageing launched since the 1950s, but it was also incompatible with the second ambition of investigating the factors influencing mortality and to thus align the BLSA with risk factor epidemiology. For this to be achieved, the initial cohort would have to be sufficiently large so as to minimise the effects of attrition and mortality in the statistical significance of observations in the years beyond middle age. Andres suggested that otherwise the study would seem to flout its mandated commitment 'to improve the quality of life for older Americans'.

The best way of ensuring this societal benefit was to curtail attrition and mortality in earlier years by recruiting the initial cohort in middle age, before the onset of disease. But as the minutes of the discussion on the day after Andres' suggestion make clear, this would necessarily render the cohort less representative of the American population generally. By selecting healthier than average individuals, who presumably would on the whole have longer life expectancies, the study would again substitute representativeness for continuity of measurement – i.e. samples for models. However, such a screening procedure could also be construed as an advantage, if the ideals of representativeness and methodological rigour were not pursued too forcefully:

> There was a divergence of opinion as to whether health screening should be imposed or not. The articulated justification for not imposing screening was that to do so would make the BLSA less representative of the U.S. population then [sic] is currently the case. On the other hand, application of such health criteria would increase the available pool of suitable subjects for many ongoing BLSA clinical studies.
>
> (NWS: 24: Baltimore Longitudinal Study of Aging Discussion Group, 18 February 1987; emphasis added)

Again, there was a tension between representativeness and continuity, which had been addressed earlier by Shock, but here it was cast as relating to the new aim of understanding the relationship between ageing and health. This focus was after all what justified the reorganisation of the BLSA onto a panel platform, for which it was necessary to ensure an 'available pool of suitable subjects'. Health screening was warranted by these infrastructural and epistemic commitments of ageing research in the new 'ageing society' assemblage. This resulted in renewing the tools for selecting a 'special, homogeneous population', a type of 'laboratory population' which, like the original recruits of the BLSA, enjoyed a healthier life than the 'average man' of their age. In sum, the BLSA could change, but not in the ways which Greulich and the NIA desired. The attempt to 'modernise' the BLSA and align it with the focus on age-related illnesses had paradoxically resulted in a reaffirmation of the uniqueness of its population within population-based approaches.

Conclusion

In this chapter, I have explored how the notion that ageing is an individualised process became entangled with a set of methodological procedures and practices encased in the longitudinal approach. I argued that this entanglement was deployed not through a process of closure but a continuously multiple enactment of the entity of the individual in the relationship between ageing, health and illness. Drawing on the history of the BLSA and the debates that have surrounded its development, I have shown that the longitudinal study of ageing has been supported by the interplay between two methodological repertoires. One repertoire has emphasised sampling and randomisation. In this repertoire, the individual is imagined as a point in a multidimensional space of statistical variations, or the combination of a series of 'deviations' from the 'average'. The other repertoire has borrowed from physiological research the notion of the 'model organism'. In this mode of knowledge making on ageing, the individual is imagined as a unique organism, whose understanding relies on comparisons with other individuals and/or itself across time points.

The institutional and political associations of the two epistemic repertoires which the history of the BLSA would seem to exemplify are worth examining. Critics of the BLSA who consistently drew attention to inadequate sampling and the need for randomisation conceived the communities examined in longitudinal studies of ageing as 'population laboratories'. In other words, they regarded ageing, illness and health as collective phenomena and as pressing problems for American society. The communities involved were thus enacted as sites which would enable access to the distribution curves and interrelations between variables within the American population more generally. On this particular understanding, sampling and randomisation enable the translation of social problems into the study of a limited, manageable set of individuals, which, in turn, secures a proportionately readier translation of findings into public policies and public health interventions in the 'real world'. On the other hand, investigators and supporters of the BLSA defined communities as 'laboratory populations'. They were interested in studying particular communities not because they were similar to any other community within the United States, but because they were usefully different – because they enabled comparison across different types of organisms. They too regarded illness as a social problem, but argued that an understanding of the precise causes of illness could only arise from comparisons with the 'pure' physiological state. This would constitute the index case that underpins modelling in the life sciences (Ankeny, 2008). Special, homogeneous communities, however much unlike any other community within the United States, offered the best approximation to such standards. This is important to the translation of any findings into public policies and public heath interventions, which is to say, to the translation back from the model to the 'real world'. Unlike samples, models in the longitudinal study of ageing called bureaucratic institutions into question by revealing how they are inadequate in their ability to attend to individual variation, to address the need to treat 'older persons as individuals, not as a class, and their wide differences in needs, desires and capacities' (see above).

My argument was that the two epistemic cultures developed in a conflictual, but co-productive, relationship. This irresolvable conflict has underpinned the transformation of the BLSA since 1958 and contributed decisively to its survival within the NIH to the present day. Confronted, in 1965, with methodological calls for representativeness, investigators at the BLSA highlighted how the selection of 'special populations' was sometimes necessary. In 1971, faced with requests to align their study with the multivariate character of ageing proposed by American gerontology, the same investigators recast the uniqueness of the population recruited into the BLSA as a variable control device. In 1978, when asked by the newly formed NIA to account for the low scientific productivity of longitudinal studies, supporters of the BLSA emphasised the epistemic and political uncertainties underpinning this negative assessment. Ten years later, when forced to become a 'panel platform' and focus on the aetiology of illness, staff at the BLSA were able to reconfigure the requests by emphasising how the uniqueness of the population recruited into the BLSA could sustain the NIA's ambition to combine the study of disease in an ageing population. At each point, challenged by statistical reasoning, gerontology, science policy and science management, supporters of the BLSA were able to reconstruct the aims of the study and its relationship to public health, the organisation of health care and the study of ageing. Overall, these episodes show the resilience of the 'model organism' agencement and the generative, interactive relationship it established with the 'sampling' approaches to explain the survival of the BLSA.

It is this internal multiplicity, I propose, that enabled the BLSA to become the central point of reference – the obligatory passage point – in establishing the view that ageing is an individualised process. In linking such multiplicity to the collective investigation of the relationship between ageing and health, particularly at the turn of the 1980s, the BLSA – its methodological approach and guiding concepts – became conditionally entwined with the problematic of the 'ageing society'. But, in so doing, it reinforced the uncertainties that relate to this problematic. In particular, it kept open the debate about procedures and instruments that should be used to investigate and manage the relationship between ageing and morbidity. Should these position individuals in a multivariate space of risk created by studies of 'population laboratories' or rely on comparisons between the individual and an 'index case'? As the BLSA transitioned from a focus on 'pure ageing' to studying health and disability, these questions gained further relevance because they questioned whether health could be understood through the study of unhealthy, real populations. Was health a Queteletian 'true value' extracted from observed variations or could it be studied directly in special, extraordinary organisms? This question, as we shall see in the next chapter, was of key importance in making healthy and active ageing the solution to the economic challenges of the ageing society.

Chapter 7

Reworking ageing

Introduction

Previously in this book, I have suggested that the 'ageing society' is deployed around a cleft problematic that relates health, activity/work and technology. In negotiating the possible solutions for the knowledge crisis of actuarial sciences, the ageing society became locked into the need to manage 'health expectancy' and productivity, and to take care of the 'existing stock' of population, rather than relying on fertility or immigration. The preceding two chapters have explored the thorny and contingent ways in which researchers have tried to measure the link between ageing and health, and the resulting increase in uncertainty about how to format health, illness and mortality. This uncertainty is underpinned by a co-productive, interactive relationship between different ways of organising knowledge making practices and institutions in managing activity and health in later life. In these, work and technology were mostly cast in a background role, or as the driving motivations for new ways of measuring or investigating ageing.

In this chapter, I focus on work and technology, and their relation to the 'ageing society'. My point of departure is, again, the epistemic and political formatting of later life through active and healthy ageing instruments and programmes. As we have seen, active ageing is the group of policies, programmes and interventions that aim to increase older people's participation in the economy, and particularly in the labour market. Within this wider context, there is a Europe-wide shared goal to postpone and increasingly harmonise retirement age across member states. As the authors of the EU *Active Ageing and Solidarity Between Generations* report acknowledge, this requires us to 'to look beyond basic measures of demographic change [...] with a focus on indicators that measure the propensity of older people to continue in work' (Eurostat, 2012: 8). But what are these indicators and instruments? My suggestion in this chapter is that 'indicators that measure the propensity of older people to continue in work' can be placed within the group of tools linked to the concept of functional age, as discussed in Chapter 5. These are tools and instruments that measure and manage individual functional abilities, and imagine a corresponding articulation

– an adjustment – to the roles or tasks s/he may be involved in. As such they deploy mainly an efficiency form of justification, proposing to maximise older people's participation in the economy by identifying unused capacities and opportunities to employ them.

In the chapter, I explore the genealogy of a particular instrument, the Work Ability Index (WAI), an indicator developed by Finnish occupational health researchers (Tuomi *et al.*, 1998), which is gaining increasing traction within work-related institutions in the EU and abroad. The WAI measures employees' individual self-assessment of their health and ability to respond to the demands of the job. My suggestion is that the WAI is underpinned by key epistemic and normative ways of enacting bodies-at-work. In formatting the instrumental value of work, functional age measurements are concerned with understanding work as a function of the internal dynamics of the body-in-action, and in particular in understanding the drivers of responses to stress and strain. It is this attention to 'internal equilibrium' and homeostasis in adaptation and the generation of stress responses that links contemporary tools, such as the WAI, to the origins of the concept of functional age, originally formulated by the industrial psychologist Ross McFarland, as indicated in Chapter 5.

The chapter details the specific conceptual, methodological, normative and institutional network that underpinned McFarland's development of the concept during World War II, and in particular his previous work in the Cambridge Psychological Laboratory and the Harvard Fatigue Laboratory. I suggest that these two sites generated different versions of the relationship between ageing and work, one underpinned by the idea of individual adaptation (Harvard) and the other concerned with identifying key mental abilities in hierarchies of working skill (Cambridge), the latter of which came to dominate the field of industrial gerontology until the 1980s. This dominance was challenged by the emergence of the post-industrial economy and what is sometimes called the post-Fordist mode of production. Normally associated with a shift in economic activity from manufacturing to services and 'knowledge-based' enterprises, for industrial psychologists this represented a vast challenge on how to design implements that are adequate for a tailored, flexible system of production. By wanting to implement ideals of 'standardised differentiation' (Busch, 2011) in the measurement of age-related abilities, experts and policy makers came to again rely on the concepts of stress and adaptation to propose solutions to the ageing workforce, thus recovering much of the work developed by McFarland at the Fatigue Laboratory.

The chapter first outlines how the economic consequences of demographic ageing have been linked to Europe's current productivity crisis, and how solutions to this problem have been consistently linked to the implementation of 'active ageing' programmes at work. Focusing on the WAI as a paradigmatic embodiment of active ageing instruments, the chapter then explores how shifts in the enactment of functional age have relied on a variation of the tension between different repertoires of knowledge making on ageing already explored in previous chapters.

Implementing the economics of ageing

At the turn of this century, marking a renewed confidence in the economic and social project of the European Union, one of the key policy groups in the EU – the Economic Policy Committee – identified demographic ageing as one of the main challenges facing European economies. As a consequence, it established the Working Group on Ageing Populations (AWG) to examine the economic and budgetary consequences of ageing. Conceived as an essential actor in the 'economic and budgetary surveillance' of member states, the AWG is composed of economists, demographers and policy makers reporting regularly on the long-term sustainability of public finances, as this is seen to be dependent on the relationship – the required transfers – between employed and retired individuals in different economic spaces. Assessing the economic consequences of population ageing for the European space as a whole, the AWG proposed that,

> Between 2012 and 2017, rising employment rates will offset the decline in the working-age population: during this period, the working-age population will start to decline as the baby-boom generation enters retirement. The ageing effect will dominate as of 2018, and both the size of the working-age population and the number of persons employed will be on a downward trajectory. [...] As a result of these employment trends [...], potential GDP growth is projected to decline in the decades to come.
>
> (Economic Policy Committee, 2006: 8–9; see also Economic Policy Committee, 2015)

In a stark evaluation of the role of demographic ageing on economic growth, the AWG identified the end of the 2010s decade as the transitional stage, dominated by the move of the baby-boom cohort – those born between 1945 and the early 1960s – towards retirement. The economic 'downward trajectory' is compounded by increases in public spending associated with the same demographic trends, particular in relation to social security and health care. Such a negative scenario is powerfully encapsulated in the assertion that the European old age dependency ratio is projected to double between 2015 and 2060, moving the EU towards 'having four working-age people for every person aged over 65 years [instead of the current situation of] about two working-age persons' (Economic Policy Committee, 2015: 1).

In projecting this 'age-related spending', the AWG are required to assume a no-policy change picture in relation to retirement entitlements, and the diverse system of financing of pensions or health care across variations on what is sometimes called the European social model of welfare. However, they are enabled to provide assessments on the likely effects of particular reforms on this 'age-related spending'. For example, in relation to retirement schemes, the AWG argued that,

One way to increase the effective exit age from the labour market in line with an increase in the statutory retirement would hence be to extend the required years of contributions or to improve incentives to stay longer on the labour market.

(Economic Policy Committee, 2015: 65; Chapter 3)

In making this suggestion, the AWG was in alignment with the policy orthodoxy of the EU and the OCDE, proposing an increase in retirement age supported by active ageing programmes. Active ageing has, as we know, been the central policy concept in the EU – and the WHO – in relation to the economic challenges of an ageing population (Lassen and Moreira, 2014). Conceiving active ageing as 'a coherent strategy to make ageing well possible in ageing societies' (European Commission, 1999: 1), the EC has proposed that this policy 'is about adjusting our life practices to the fact that we live longer and are more resourceful and in better health'; 'in practice, it means adopting healthy life styles, working longer, retiring later and being active after retirement' (European Commission, 1999: 2). In the intervening years, the EU active ageing policy has become an amassed, evolving cluster of pension reforms, age integration policies, and age management programmes. Such changes in the European attitude towards labour market withdrawal and 'old age' represent for some commentators 'the unusual combination of a morally correct policy that also makes sound economic sense' (Walker, 2002: 1). However, as Avramov and Maskova (2003) recognise, as a set of value and norms active ageing is difficult to translate into concrete policies and to be accepted by the majority of the population. For this reason, in the past decade increasing numbers of actors have been arguing that bureaucratic changes in pension entitlement age fall short of addressing the width and depth of the dimensions underpinning processes leading to labour market withdrawal and transition to retirement.

For example, the European Network for Workplace Health Promotion – an informal network of national occupational safety and health organisations and actors in the field of public health – recognising that 'working life has not been extended as wished' in a number of EU countries, proposed that 'the most important force for change is the workplace' (Morschhäuser and Sochert, 2007: 6–7). Working in close partnership with the WHO European Regional Office and the European Commission, this network of experts proposed that health and the perceived effects of work on wellbeing were the key drivers of withdrawal from the labour market, there being no consistent evidential basis to use chronological age as an index for efficiency or productivity. Indeed, as a latter review of the literature on age and employment for the UK Health and Safety Executive put it, when abilities match job requirements and when experience is taken into account, 'there is little difference between the performance of older and younger workers' (Yeomans, 2011: vii).

In establishing this consensus around the role of the workplace on health and on the adjustment between abilities and tasks – which is dear to most experts on age and employment – one particular approach to the understanding of age-related work processes can be identified as decisive. This is an approach to managing ageing and retirement mostly identified with Finland. Affected by an economic recession in the 1990s, and facing a projected transition to an age-dominated labour market about ten years earlier than the rest of the EU, Finland embarked, from 1998 onwards, on a series of policy initiatives that shaped the 'active ageing agenda', not least because the foundational 1999 Active Ageing Conference, refereed above, was held during the Finnish presidency of the EU. The success of this series of initiatives was enacted by coordinated and inclusive policy processes supported by a strong knowledge base provided by the Finnish Institute of Occupational Health (Piekkola, 2004). This was mostly drawn from the findings and models used in the 'Finnish research project on ageing workers in municipal occupations' (Tuomi et al., 1997), which deployed the WAI as its main instrument of functional age. In this, the WAI was seen to embody Finland's innovative approach to retirement – which will become enshrined in the Finnish pension system from 2017 onwards – in that it recognises the individual, health-related variations in ability to work at retirement age.

In its simplest definition, the WAI is a scale resulting from a self-administered questionnaire focusing on the respondent's assessment of changes in work ability, health and present ability to respond to the demands of the job. Calculating these, as well the respondent's own prognosis of years until retirement, the cumulative index of the WAI ranges from 7 to 49 points, divided into the categories of poor (7–27), moderate (28–36), good (37–43) and excellent work ability (44–49). The WAI score is said to predict the probability that a person will retire early (Tuomi et al., 1997). In this capacity, it can be used as an instrument for assessing the dynamic of working age populations across countries, sectors or occupations. In addition, the WAI can also be used at the level of the organisation, by individual employees (Ilmarinen, 2010; Ilmarinen, 2006). In this regard, the WAI serves as a human resources planning tool for companies, and in supporting workers in managing their transition to retirement. This is of key importance, because, in this respect, the WAI is both a measurement and a tool for action, identifying whether to maintain, support, improve or reinstate work ability at an individual level, and setting forth a series of interventions (Weyman et al., 2012). It is a pragmatic instrument. In this respect, the WAI embodies a political script that is fully aligned with the active ageing project, as it 'depicts the ability to work from a positive point of view of health and mental resources' (Tuomi et al., 1997: 7). This alignment with what we might call the active ageing assemblage is not a coincidence. Indeed, one might argue that active ageing instruments are the policy articulation of a way of knowing and valuing the economic worth of older people that has its origins in the inter-war years, as explored in Chapters 4 and 5.

Further clues to this genealogical trace can be found within reports and papers produced by the 'Finnish research project on aging workers in municipal occupations' from the early 1990s onwards. Contextualising the research interest in the wider challenges faced by the Finnish manufacturing sectors in the 1970s through the introduction of microelectronics, Ilmarinen and his colleagues viewed 'technological change' as the main driver behind the setting up of the project in 1981. As they themselves put it, 'worklife [has changed] mostly because of the implementation of new technology' (Ilmarinen *et al.*, 1991: 8), affecting the organisation of work, the nature of work tasks and the skills needed to perform those tasks. This technological problem combined with uncertainty relating to the adequacy of the Finnish occupation-dependent retirement age – established in 1964 – to buttress a re-examination of the relationship between ability and age.

The project was thus initiated in the Institute of Occupational Health in 1981 to test the use of health and capacity related criteria for determining retirement. The study encompassed a questionnaire and a laboratory component used both at baseline and in the follow-up. It specifically aimed to relate the analysis of work tasks to perceptions of changes in performing those tasks, and the physiological measurement of key functions such as cardiorespiratory or musculoskeletal capacities (Ilmarinen *et al.*, 1991). For Ilmarinen, Tuomi and their colleagues, the focus on those relationships was underpinned by a model which they labelled the 'stress–strain concept'. As they explain,

> The strain experienced by the worker depends on work stressors and also on the individual characteristics and abilities of the worker [...]. If the work demands exceed the qualifications of the worker, overstrain is induced; if they are lower than the worker's resources, understrain will follow. Both conditions can also be the result of unsuitable work in relation to individual characteristics[.] Strain may have either positive or harmful consequences. Harmful strain [...] leads to acute or chronic physiological responses, psychological reactions [...] and eventually to work-related diseases [.] In contrast, a positive combination of work demands and individual human abilities promotes health and work ability at any age.
>
> (Ilmarinen *et al.*, 1991: 9)

Rather than locating strain solely on the demands of the task or the abilities of the worker, the stress–strain concept focuses on the interactive dimensions of the relationship between these two elements. Strain is thus conceptualised as an integral, internal function in the dynamics of work, producing positive feedback loops when there is a fit between task demands and abilities, and producing harm 'if the work demands exceed the qualifications of the worker'. Explicitly linking this approach to the German tradition of occupational medicine (*Arbeitsmedizin*), the project was also emplaced within a wider problematic of stress at work that has permeated ergonomics, industrial psychology and occupational health research over the past century. In particular, as can be seen in the quote provided above, it was affiliated to theories of stress that focus on the person–environment system, and which concentrate on the degree of match or mismatch between the individual and his or her environment (e.g. French *et al.*, 1982). In this approach, 'stress is neither in the environment nor in the person' (Lazarus, 1990: 3), being instead defined as the function of a relationship in which there is a deficit of resources available to meet the demands that the relationship requires.

In tracing the stabilisation of the concept of stress as a central problematic in the study of work, researchers in the field, including those in the Finnish study of ageing workers (Tuomi *et al.*, 1997), and historians of the human sciences would customarily refer to the work of Hans Selye. As Viner (1999) and Jackson (2012) have suggested, however, while Selye did much to link his name to the collective recognition of the importance of stress for the modern human condition, his conceptualisation of stress was itself embedded in shared forms of reasoning about self-organisation in living organisms and thinking machines. Indeed, both in Selye's original laboratory work on the 'general adaptation syndrome' and stress in rats and his subsequent proselytising on behalf of the general applicability of the concept, he recognised his indebtedness to the work of Bernard on the '*milieu intérieur*' and Walter Cannon's related idea of homeostasis – physiological equilibrium. Similarly, while Selye did not make

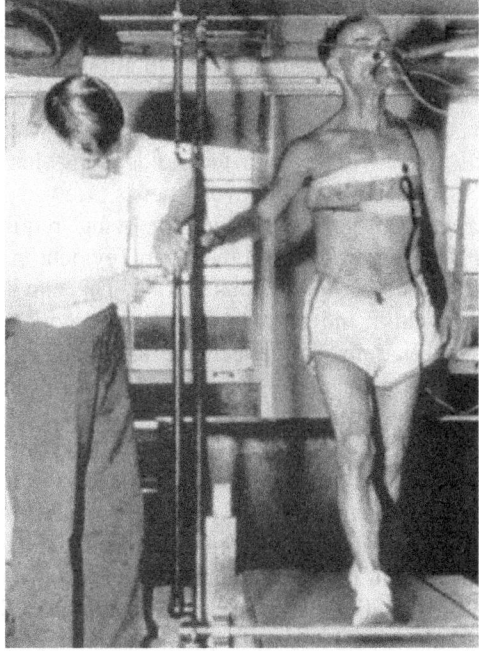

Figure 7.1 Clarence DeMar walking on the Harvard Fatigue Laboratory treadmill in August 1937, at the age of 49. Observations were being made on his heart rate and respiratory minute volume. Sid Robinson, who at this time was a graduate student at Harvard, is shown obtaining a blood sample for lactate determination.

direct reference to the cybernetic research on feedback and self-organising machines, 'he was clearly aware of systems philosophy' that stemmed from it (Jackson, 2012: 24). Selye's own realisation of the overlaps between these two fields of knowledge in supporting the idea of stress is, from my perspective, indicative of established 'circuits of translation', rather than mere intertextual connections underpinned by widespread cultural beliefs. To make those circuits of translation visible, however, it is necessary to depart from practitioners' histories of occupational health research and of the WAI itself, and suggest an alternative genealogy that follows the tools associated with the concept of functional age, which will be the concern of the next section.

Ageing and efficiency I

In Chapter 5, I have suggested that the consolidation of an epistemic and normative critique of chronological age as a criterion for the distribution of rights and responsibilities in 'old age' in the inter-war years was associated with the articulation of alternative, substitute forms of age measurement. I have argued that these alternative age standards proposed a form of 'standardised differentiation' that aimed to conjoin individuals' unique capacities or needs to roles or services. I identified four broad categories of substitute age measurement and suggested that those linking a focus on functional capacities to an orientation towards efficiency shared key epistemic, political and institutional elements organised around a concern with the effect of technology on work. Sharing Lawrence Frank's negative evaluation of the impact of modern technology on what he called the organism-personality (see Chapter 2), psychologists viewed the concept of functional age as underpinning the new 'designs for living' required by the new world of industrialised work, a commitment that is still evident in the origins of the WAI instrument, as explored in the previous section. This aim supported much of the work developed within industrial gerontology, ergonomics and human factors engineering on the relationship between age, skill and work.

One of the key figures in the establishment of these fields and their reformist orientation towards work, as I also suggested in Chapter 5, was Ross McFarland (1901–1976). McFarland can, with some certainty, be seen as the originator of the idea of functional age. In this, his proposal was to bring functional age to fruition by deploying 'not only more accurate aptitude tests for placement of the older worker, but also more refined methods of evaluating his ability to respond to fixed stresses' (McFarland, 1943: 509). Such focus on stress and adaptation was distinctive because it questioned the widespread belief that there was a fixed relationship between age and functional capacity. Functional age represented thus an important tool in reassessing the value of older people's labour by focusing on the technology–human interface, and the dis-adjustments – *the stress* – it produced. In proposing this stress-focused programme of research and action, McFarland was drawing on his experience of working in two laboratories: the Cambridge Experimental Psychology Laboratory and the Harvard Fatigue Laboratory.

Led by Frederic Bartlett (see Chapter 5), the Cambridge Experimental Psychology Laboratory was by 1927, when McFarland joined it as a post-doc, involved in a programme of research that investigated the effects of altitude and oxygen deprivation on the performance of aeroplane pilots. This was linked to a set of academic activities which linked the expertise of psychologists to the military. Such activities had supported a close collaboration between the Cambridge Laboratory and the Armed Forces since World War I, a collaboration intimately hinged on using the laboratory to configure the relationship between modern warfare technology and the mind of the soldier. Believing it was the business of the psychologist to understand the *human mechanism* which 'all the numerous [warfare] applications of physics, chemistry and engineering are at the mercy of' (Bartlett, 1927: 2), Bartlett was especially concerned with using the laboratory to test the match or fit between military designs and human capacities.

With the advent of what has been sometimes called the 'golden age of aviation' – marked by the first transatlantic flights in 1919, and the transition to fast all-metal monoplanes – the Cambridge Laboratory was asked to investigate the effects of increased altitude and flight time in the performance of aviators. Investigating the fit between these new technologies and the *human mechanism* entailed, however, extending the collaborative network to include the expertise and experimental approach used by physiologists, and in particular that of Joseph Barcroft. Best known for his studies of the oxygenation of the blood (Barcroft, 1914), Barcroft had been leading a research programme on the physiology of oxygenation at extreme altitudes, organising for this purpose expeditions to the peak of Tenerife, Monte Rosa and the Peruvian Andes. It was in the context of this collaboration between Bartlett and Bancroft that MacFarland first came into contact with the role of oxygenation in cognition. Using Royal Air Force subjects, McFarland used his time at Cambridge to investigate the effects of lack of oxygen on the cognitive performance of pilots, realising the key importance of progressive exposure and 'acclimatisation' in this relationship.

Returning to the US, McFarland's research became focused primarily on the development of a technological environment – the pressurised cabin – where pilots could offset the consequences – the physiological stress – incurred in rapid ascent. Funded by the US Department of Commerce, these investigations likened the effects of anoxia – lack of oxygen – to the symptom of psychiatric disorders such as 'dementia praecox' or psychosis. This connection and the recognition of its significance for the modern, technologically intensive workplace moved McFarland, in the years between the wars, to a central position in the development of what later became known as 'human factors' research. This central position was sealed with McFarland's move to the Harvard Fatigue Laboratory in 1937, to investigate the effects of commercial transatlantic and transpacific flights on pilots' continued performance. Commissioned and generously funded by Pan American Airways, it provided McFarland with a focus and a sustained source of research subjects (training pilots), actively contributing to the Laboratory's programme of work.

The Harvard Fatigue Laboratory had been established a decade earlier as a collaboration between Lawrence J. Henderson, a physiologist, and Elton Mayo, a psychologist of labour relations. In his preceding work, Henderson had advocated an integration of Bernard's idea of '*milieu intérieur*' within a wider conceptual and instrumental assemblage, with an emphasis on the capacity for organisms to adapt to their external environment (Henderson, 1917; Parascandola, 1971). Rejecting mechanistic understandings of physiological processes, Henderson drew on the work of Italian sociologist Pareto to conceive of internal equilibrium as a dynamic, multifactorial, interdependent state, resulting from the interaction of 'mutual dependent variables' (Cot, 2011). This approach to organisms and society was crucial in gathering support and funding for the laboratory because it presented a neat solution to one of the key labour problems of the inter-war years. This was related to the political corollaries of an understanding of fatigue which drew mainly on the analogy between the body and heat engines, hinging on the principles of thermodynamics (Rabinach, 1990). Such an enactment of human physiology at work buttressed an alignment between fatigue research and progressive labour reform, which motivated policy calls for limiting work time and pace. These, however, came under attack within the context of the US and European establishments' reactions to the formation of the Soviet Union in the 1920s (Gillespie, 1987).

In accruing funds from the Rockefeller Foundation to set up the Fatigue Laboratory, Henderson and Mayo aimed explicitly to redress this thermodynamic-based, dominant view of human labour, and alternatively tie fatigue research 'to business rather than reform' (Scheffler, 2015: 399). Claiming that 'next to nothing' was known of the effects of the new industrial environment in the activities and mental processes of workers, Henderson envisaged the Fatigue Laboratory as a new departure in this field of research by 'studying the mutual dependence of quantitatively large physiological activities in different normal men [...] to make trustworthy quantitative distinctions and discriminations between individuals' (Henderson in Scheffler, 2015: 405). This aim supported them in a series of experiments that attempted to measure physiological responses to controlled amounts of physical activities in different individuals, from which the exercise treadmill apparatus emerged as an instrument for research in industrial physiology, and exercise science more generally, until the present day (Scheffler, 2011). Comparing the responses of athletes – as model organisms – such as marathon runner Clarence de Mar with those of 'normal men' (see Figure 7.1), Henderson and his colleagues identified a relationship between training – adaptation to stress – and performance. Most significantly, in comparing the lactic acid increase – a contemporary marker of fatigue – in response to exercise, Fatigue Laboratory researchers had found that those who had never trained for running experienced the highest levels of lactate increase.

Translating these findings to those interested in industrial relations, Mayo suggested that fatigue should be viewed as a '*maladaptation*' between the 'external condition and the individual' (Mayo, 1933: 22–23). Thus, contrary to the then dominant view on bodies-at-work, the problem of fatigue was not under-pinned by the pace of industrial labour or the abilities of the 'average worker' but by their relationship; and its solution, provided by industrial physiologists and psychologists, was to engineer the *fit* between the worker's abilities and the task demands. This recasting or questioning of the very concept of fatigue hinged on the ability to materialise and make visible through instruments such as the treadmill the differential, individual ability to respond to a controlled source of stress. It was compounded by the manipulation of the 'mutual dependent vari-ables' in experimental environments such as high-altitude chambers or 'cold rooms', stabilising a collective investment around the ability of physiologists or psychologists to re-establish the 'internal equilibrium' even in the most adverse of work situations.

Interestingly, from our perspective, in analysing the graphs of differential lactic acid increases of research subjects, Mayo observed, in his *The Human Problems of an Industrial Organization*, that the 'internal equilibrium' was a function of previous training in running rather than age, noting,

> with respect to the group of persons involved as subjects in this experiment, that the older men [experience less lactic acid increase]. It happens that the best in the group, the athlete [De Mar] was forty at the time of the experi-ment; the worst, at the extreme right, was a boy of eighteen. This does not lead, of course, to consideration of the effect of age, but leads rather to con-sideration of the effects of physical training.
>
> (Mayo, 1933: 14)

That Mayo made this observation is indicative of the significance of age and 'age effects' in the recasting of the concept of fatigue. Indeed, Mayo's sugges-tion is aligned with the arguments explored in Chapter 5 of this book in that he uses physiological measures to undermine the presumed effects of chronological age in human performance. While he was doing it primarily to show that the 'ability to maintain an internal environment' dominated the presumed effects of ageing in the organism, it also opened an angle of research for the industrial physiologist in the emerging field of gerontology of the 1930s, an angle that was fully aligned with Cowdry's and Frank's vision for gerontological research. It was only with the US' integration into the Allied forces in World War II in 1941, however, that this possibility came to fruition.

The mobilisation of Fatigue Laboratory staff to the war effort started in fact in the previous year, and focused on research on topics such as the effect of alti-tude and temperature on military personnel (Folk, 2010). Not surprisingly, given his expertise, McFarland was part of that collaboration establishing, in 1940, the Pensacola Study of Naval Aviators (aka 1000 Aviator Study) to develop better physiological and psychological tests for predicting success in pilot training. This would be followed by studies of air and ground force fatigue, which, com-bined with his studies of commercial aviators, would provide the basis for his book *Human Factors in Air Transport Design*. In this book, seen by many as the seminal foundation of 'human factors research', he argued, in line with Henderson and Mayo, that understanding adaptive responses to stress at work would lead to a 'more effective integration of men and machines' (McFarland, 1946: 46).

His involvement in the war effort also included a small study – what we would now call a narrative literature review – of the employment of the 'older worker in industry' (McFarland, 1943). In this context, he was able to fully apply the Fatigue Laboratory's stance to the 'problems of ageing'. The study, commissioned by the Air Force, aimed to understand the possible effects of employing 'retired workers' on production tasks in the aircraft manufacturing companies of Southern California, faced with the combination of increased demand and decreasing supplies of labour caused by conscription. Contextualising the need for the study beyond the war situation, in the wider process of demographic ageing, McFarland identified 'prejudice' against older people and the lack of 'scientific research' on ageing and performance as the main obstacle for employing older workers in industry. Challenging these uninformed policies and practices, McFarland argued that,

> Growing up and growing old are continuous processes and many changes occur throughout the lifespan. But changes with age do not necessarily mean decline. Compensation takes place for every deviation, and if certain capacities diminish, others are enhanced. For example, as speed of reaction is lowered with age, there occurs a compensatory increase in endurance. This fact has been revealed in many different ways. In athletic performance there is a positive correlation between maturity and success in competition requiring endurance. [...] Some men even at 50, such as Clarence de Mar, continue to show outstanding performance in marathon races. Furthermore, greater differences are observed in an exercise endurance test in persons of the same age group than are observed between younger and middle-age groups.
>
> (McFarland, 1943: 507)

Conceiving of changes in the 'internal environment' as 'mutual dependent variables', McFarland suggested that fluctuations associated with ageing are embedded in a dynamic equilibrium where 'compensation takes place with every deviation'. To demonstrate this, he, like Mayo, drew on the figure of Clarence de Mar and the treadmill laboratory experiments. These, and his own experiments on the 'response of older persons to a fixed stress in the field of aviation', enabled him to claim that chronological age or age grouping could not underpin what Henderson viewed as 'trustworthy quantitative distinctions and discriminations between individuals'. It was necessary to invent new metrics that would support the effective integration – *the fit* – between older men and machines in industry:

Osler said that 'a man is as old as his arteries'. In industry, it would be equally sound to say that a worker is as old as his vision, his motor skill or his productivity. The important factors to consider in employing older men relate, not to chronological age, but to 'functional age' or ability to perform efficiently the tasks involved in specific jobs. The great need in industry is, not only more accurate aptitude tests for placement of the older worker, but also more refined methods of evaluating his ability to respond to fixed stresses.

(McFarland, 1943: 509)

The use of quotation marks around the term functional age is indicative of the uncertainty with which McFarland was proposing it in 1943. Contrasted with the more established measurement of chronological age, or the emerging concept of 'physiological age', the new term encapsulated the new approach to the application of physiology and psychology in industry proposed by Mayo, Henderson and McFarland. This approach entailed not only the use of aptitude tests, designed to emulate the content of specific tasks but also, and importantly, laboratory investigations of the response of workers to 'fixed stresses'. This articulation of the psychological test with the experimental setting of the questionnaire and the laboratory, which was to become central to industrial psychology – and later on to the occupational health research and practice embodied in the WAI – constituted the epistemic anchorage of the concept of 'functional age'. For McFarland, functional age should work as a predictive tool, assisting the industrial physiologist in finding an individualised fit between workers' abilities and work tasks. Furthermore, the identification of workers' ability to respond to stresses (his 'physiological and mental reserve') should 'lay the foundation for effective preventative medicine' (McFarland, 1943: 510; see also McFarland, 1956).

Ageing and efficiency II

MacFarland's proposed arrangement of expertise and instrumentation in the management of labour relations was brought to bear by the combination of the effects of the 1935 Social Security Act and specific circumstances of warfare manufacturing in World War II. This combination was somewhat ameliorated in the postwar years in the US, until the debates around the creation of Medicare brought it again to the fore. In the United Kingdom, by contrast, the post-war years saw the establishment of non-contributory old age pensions and the effects of war mortality on the structure of the working population (see Chapter 3). These set the context for a renewed interest in research on the employment of older people in industry. No organisation was in a better position to address this need than the Cambridge Psychological Laboratory, which since McFarland's departure had further reinforced its collaboration with the Armed Forces. This was embodied in the establishment of the Unit for Research in Applied Psychology in 1944, supported by the Medical Research Council and the Rockefeller Foundation. Highlighting its distinctive experimental approach to 'human skilled performance', the Cambridge

Laboratory seized the new opportunities related to the 'problems of ageing' and obtained funding from the Nuffield Foundation to establish the Research Unit into Problems of Ageing at the Cambridge Laboratory in 1946.

It is worth exploring the approach of the Nuffield Unit because it represented, at the time, an epistemic and political alternative to the Fatigue Laboratory. While the Fatigue Laboratory's work hinged on Henderson's physiological concept of 'internal equilibrium' and its embodiment in the exercise treadmill, the Nuffield Unit's work was 'mainly psychological in orientation, dealing with mental capacities and skills [albeit] recognising the need to keep in touch with physiological and medical studies' done elsewhere (Welford, 1966: 2). Key to this discipline-based approach was the leadership of Frederick Bartlett. In the 1930s, Bartlett had developed a psychological theory of memory that suggested an active role of 'schemata' in constructing and organising remembering (Bartlett, 1932). The schemata were conceived as organising principles, actively selecting received impulses, themselves dynamically changing as the needs and situation of the organism developed, 'producing an orientation of the organism towards whatever it is directed to at the moment' (Bartlett, 1932: 208). Like Henderson, Bartlett drew on the social sciences – anthropology in his case – to challenge mechanistic, 'reflex' theories of remembering, arguing instead that it relied on dynamic self-organising structures.

But while Henderson conceived of self-organisation as the continuously changing configuration of 'mutual dependent variables', for Bartlett this relied on threshold conditions that changed the shape of the schemata or the relationship between them. The significance of these *keys* was brought to bear when Bartlett was asked by the Air Force to investigate fatigue in pilots during World War II. Wanting to avoid replicating what he saw as the 'non-sense' of much of classical psychology's use of experimental designs, detached as they were from real-life situations, Bartlett asked his assistant Kenneth Craig to devise an experimental cockpit. Recalling a few years later how this experimental device changed their approach to skilled performance, Bartlett noted:

> It is fair to claim that these experiments, more than any others, gave us the guiding principles which we have been able to use, in many applications, with growing definiteness, ever since. [...] We knew now that in order to discover anything important about skilled performance we must find a way to get right inside the skill itself, and to relate, within the experiment, actual overall measures with actual measures of the ways in which the internal constituents were being carried out. It became plain that, certainly in this type of skilled performance, and perhaps in others, not all of the internal constituents are equally important. There are 'key' or 'master' elements round which all others tend to be grouped, so that if the key activities are properly timed and carried out, all the rest will normally go well without much, if any, voluntary effort.
>
> (Bartlett, 1951: 210)

The Cambridge Cockpit, as the experimental device came to be known in later years, had the same impact in stabilising Bartlett and colleagues' view on skill as the exercise treadmill had on the Fatigue Laboratory's approach. It enabled the Cambridge team to 'get right inside skill itself' by making visible a hierarchy of 'internal constituents' where certain activities dominate and organise the performance of other, lower elements. Those key or master activities had the function of controlling a cluster of elements which required less conscious focus, resembling the system of inhibitory controls which Cambridge electrophysiologists had described a few years earlier (Smith, 1992). From this perspective, there was a parallel between the structure of the controls of the aeroplane – the machine – and the organisation of what Bartlett had once called the 'human mechanism' (see above). This parallel in part justifies considering the Cambridge Cockpit as a precursor to cybernetics (Adey, 2010), in that the epistemic equivalence of man and machine enables the conceptualisation of feedback relations, supporting self-organisation.

The Cambridge Cockpit was also pivotal for configuring the approach to research in skill and age carried out at the Nuffield Unit. In particular, it justified a move away from single 'stimulus-response' assessment of age effects towards experimental designs that 'get inside skills' by respecting their 'natural order of expression' (Bartlett, 1951: 215). This was, however, a problem because while in studying aeroplane flying there was some stability and uniformity in technological design, industrial workers were faced with many types of machines, with new ones being introduced, at the time, with electronic components. Choosing to focus on the 'relatively light skilled operation' in which they had most expertise, and which they saw as the future of industry, the Nuffield Unit, led by A. T. Welford, focused on identifying the *key or master elements* that affected skill in old age.

Such a focus on the key elements determining the speed, accuracy and general efficiency of the older worker marks another crucial difference in relation to the Fatigue Laboratory. Bartlett and Welford's experimental designs were aimed at bringing to bear the organising principles affecting skill with age (Welford, 1958). In this respect, they were not geared towards differentiating organisms, but to explaining why organisms, *on average*, changed in a particular way. Key elements, like schemata, pertained to a higher level of epistemological and ontological significance than organisms conceived as a whole dynamic unit. We have here, again, a version of the divergent ways of enacting age already found in the previous two chapters: one based on model organisms and the comparison between individuals and the 'index case' (Harvard), and the other based on the extraction of 'true values' – key elements – from observed instances of skill in ageing individuals. In this context, one used the laboratory to identify the simple, pure relationships between variables in model organisms, while the other used the laboratory to generate variations within populations.

These epistemic commitments were in close relation with the normative and political position articulated by the Nuffield Unit in relation to age grading of work responsibilities. Instead of a full endorsement of an individualistic position, encapsulated in the concept of functional age as proposed by McFarland, Welford and his colleagues argued for 'more flexible' policies of retirement (Welford, 1958). In this regard, Bartlett and Welford were politically not far away from the solutions proposed by Lord Beveridge (1943) himself. Recognising that 'individuals seem to change with age at different rates' (see Chapter 6), Welford however added that,

> the abler tend to change more slowly than the less gifted, so that variation between individuals increases as they move up the age scale. The obviously important implication is that policies for older people must neither be based on exceptionally able individuals nor, on the other hand, fail to make provision for them.
>
> (Welford, 1966: 5)

For Welford, variations in skill with age were a function of ability, particularly those related to the 'higher functions in the brain'. Conceding that policies could not solely provide for 'exceptionally able individuals', he was concerned that universal policies failed however to recognise them. This meant that talent and ability were wasted. However, for him, the role of the industrial psychologist was mostly concerned with those close to the average of the normal curve of ability, or as speaking to the needs of the 'normal' industrial worker. As he put it, the role of the industrial psychologist in relation to ageing was 'to specify work which is suitable for the majority of older people' (Welford, 1953: 1194). Adjustments of job tasks and pace were thus consistently related to age groupings rather than individual differences, such that 'a normally healthy population should be able to maintain and modify acquired skills [...] well beyond any age that is now considered to mark the term of an active working life' (Bartlett, 1951: 516).

Thus, in light of the arguments presented in Chapter 5, it is understandable that Nuffield Unit researchers never explicitly advocated policies hinged on the concept of functional age (but see Welford, 1965), as they were not aiming to differentiate individuals on the basis of 'the ability to perform efficiently the tasks involved in specific jobs' (McFarland, 1943). Instead, drawing on population variations, Nuffield researchers were committed to identifying the generalised drivers – the *key elements* – in skill that change with age. Knowledge of these, they believed, should be the source of a capacity to intervene in organised industry so as to 'heal troubled bodies, troubled minds, and a troubled society' (Bartlett, 1951: 217). This represented, as can easily be seen, a different solution to the efficiency and 'human problems in industrial civilization' advocated by Mayo and his colleagues at the Fatigue Laboratory, aiming to cluster, reclassify and regroup older people according to skill-based standards rather than to discriminate between individual cases.

The differences between the Nuffield Unit and the Fatigue Laboratory are redolent, as I have suggested above, of the conflict between Shock and 'the NIH statisticians' discussed in Chapter 6, opposing advocates who were speaking either for the representative majority expressed in samples or for the case as a model of 'pure ageing'. However, the divergences here were never expressed through open controversy, even allowing McFarland and Welford to collaborate, for example, on research on automobile driving and ageing (McFarland et al., 1964). Instead, what we observe in relation to different enactments of age and skill is a form of distribution between the two approaches (Mol, 2002), whereby McFarland's concept became incorporated within the psychology-driven, variable-based research field known as industrial gerontology between the 1950s and the 1970s. Evidence of this inclusion can be found for example in the methodology proposed by Dirken (1972) or Heron and Chown (1967) for the measurement of 'functional age', where McFarland's original concept is used to describe profiling measures of decline on specific behavioural variables (Schaie, 1977). Indeed, in an important review of the concept in the 1980s, Salthouse could observe that while the origins of the concept were inextricably tied to McFarland, most research on the concept focused on the 'development of a battery of abstract measures' applicable to a variety of jobs (Salthouse, 1986: 80). Such 'abstract measures' were, of course, the 'key elements' first made available through the Cambridge Cockpit experiments.

The end and beginning of 'functional age'

The dominance of the Cambridge approach over industrial psychology's enactment of the problem of ageing and skill was particularly evident between the 1950s and the 1970s, a period where consistent economic growth combined with an increased recognition of the value of psychology within the 'world of work'. This world had been shaped by the economic and policy arrangements of the post-war era, particularly its focus on the drivers of efficiency in 'industrial organisations'. During the 1970s, however, these tight alignments of knowledge and work became increasingly at odds with changes in the economic framing of efficiency. This disalignment involved five processes.

First, the shift from manufacturing towards service-based economic activity between the 1970s and 1990s. This entailed a variety of economic and social transformations. In 1960, in the United Kingdom, about 35 per cent of the working population were employed in manufacturing; by the year 2000, there were less than 15 per cent. Similar trends were observed in countries such as the USA, France and Germany (OECD Labour Force Statistics). The transition to a service-based economy was accompanied, in the UK, by increased levels of unemployment, social and political conflict and the decline of particular ways of living and feeling. This represented a challenge to industrial gerontology, because, as the very name indicates, it was tied to this techno-economic and cultural arrangement of institutions and practices. This was true especially of research focused on 'batteries of abstract measures', as the types of skills and abilities required of machine operators (speed, accuracy, etc.) were largely absent from newer service sector jobs. As these variables had been abstracted from field observations of work performed by the 'majority' of workers in industrial societies (Welford, 1958), their applicability to the new service economy remained uncertain at best.

Second, while the post-war period, particularly in Europe, had seen the need to explore the possibility of employing older workers, in the 1980s and 1990s, a variety of policies and financial incentives encouraged a growing proportion of the workforce to retire or leave the labour market early. These were linked to the perceived need to facilitate shifts within the labour force from industrial to service types of employment, as referred to above. Aimed at offsetting the consequences of a lasting fall in employer demand for low-skilled industrial workers, which particularly affected older generations, these policies coincided with shifts in 'cultures of ageing' that extended leisure practices to sectors of the population which had been previously excluded. In this context, without a stable political scaffold, the embedding of industrial gerontologists in policy making networks was considerably weakened. As questions of transitions to retirement and its impact on workers became more pressing, psychologists and ergonomists had to extend their gaze and experimental approach to leisure and the post work world (see Chapter 8).

Third, in those same decades, in Europe and the US, we saw the consolidation of what Laslett (1989) labelled the 'third age' to describe new demographic, normative and institutional arrangements supporting positive engagement with post work life. As people enjoyed healthier later lives, and had increasing, but differing, access to economic, social and cultural capitals, there was a diversification of the experience of later life across generational, class and gender boundaries. Associated with the processes of de-standardisation or de-institutionalisation of the life course discussed in Chapter 5, such transformations established new forms of identity in later life, where consumption and leisure and the enactment of taste dominated work-related identity. In a context where work is devalued and later life comes to represent an individualised, 'reflexive opportunity' to reinvent social identity and the self (Biggs, 1997), there is clearly no need for the application of 'abstract measures' applicable to the 'majority of workers'.

Fourth, shifts in the socio-economic organisation were linked to a fragmentation of work regulatory practices, sometimes associated with post-Fordist systems of production, and the application of tools such as total quality management, lean manufacturing or just-in-time delivery. This often relied on a diversification of investments by companies, their dependence on networks of contractors spread around the world and a segmentation of markets around products increasingly 'qualified' by a combination of quantitative techniques and behavioural sciences (Cochoy, 1998). As we saw in Chapter 5, this has been connected to a shift from generalising standards and tools to those that aim to expertly differentiate preferences, abilities, needs, etc., in the workplace. With an increasingly differentiated and flexible workforce, profiles of aggregated, 'abstract measures' of skill became less useful for personnel managers and human resources staff. In this newly configured workplace, as Salthouse has recognised, 'knowledge of the average [functional] age of a pool of applicants may be helpful to predict the proportion of applicants achieving a criterion score, but this information is more relevant at the individual level' (Salthouse, 1986: 86).

Finally, the very notion of 'functional age' that relied on 'abstract measures' and the identification of 'the key elements of skill' came under critical technical scrutiny. The issue was that the aim to develop a universal battery of tests went against established knowledge about the heterogeneity of skill changes with ageing (Salthouse, 1986) and the fact that individuals' rate of ageing was unlikely to be constant through their life span (see Chapter 6). Indeed, the very construction of such ageing measurement tools 'require[d] acceptance of the irreversible decrement model of ageing' (Schaie, 1977: 4), as they relied on the establishment of age functions through multivariate or factor analysis. This reliance on regressions to the mean to gain validity made such exercises paradoxical in nature, being only able to replace one set of normative assumptions linked to age groups with another. This led prominent researchers in the field to suggest that it might be wise to rely again on chronological age as a surrogate marker of the ageing process rather than trust the promise embodied in the idea of functional age (Chapter 5).

Devising a way out of this sociotechnical predicament entailed forsaking the concept of 'functional age' to embrace that of 'work capacity'. Defined as an aggregate of 'the capacities necessary to perform a given type of work' (WHO, 1993: 3), this concept encapsulates, as we shall see, a reversal of the inclusion relationship previously observed between the Nuffield Unit and the Fatigue Laboratory's approaches to ageing and skill. This is mostly evident in the fact that the definition of 'work capacity' shares the same epistemic and normative scaffolds as McFarland's original formulation of 'functional age', in that it subsumes psychological research on skill and abilities within a wider articulation of fitness and health of the worker. This linkage between work capacity and McFarland's earlier work, the consequences of which we are still experiencing today, was first drawn by the knowledge and political work conducted at the WHO Aging and Work Capacity working group meeting in Helsinki in December 1991.

Held at the Finnish Institute of Occupational Health, this meeting is usually seen as the moment when the concept of 'work capacity' gained recognition. Including public and occupational health officials from both developed and developing countries, as well as representatives from international labour organisations, and experts in the fields of gerontology and geriatrics, the meeting aimed to 'analyse the changes in work capacity due to aging'. However, instead of dedicating the main bulk of their work, as one would expect, to revising and discussing psychological work on skill and age, the group chose instead to focus on considering the 'biological background of ageing' and how health and health promotion affected 'capacity for *adapting to work environments*' (WHO, 1993: 1, my emphasis).

This approach was justified, in their view, by the fact that it was difficult to distinguish between the effects of age and illness on work capacity. To understand the significance of this stance, it is perhaps important to remember that most research on industrial gerontology had left the issue of illness and the health of the worker relatively untouched, considering it to be within the occupational doctors' or physiologists' realm of expertise. Welford, for example, had suggested that health alone could not explain withdrawal from the labour market, as '[m]any who are successful in employment and enjoy their work are far less fit in the medical sense than many who give reasons of health for being unemployed' (Welford, 1966: 8). Challenging this consensus, the members of the WHO Aging and Work Capacity working group, led by Ilmarinen, emphasised the uncertainties regarding the accuracy and validity of the definition of fitness in the 'medical sense', because the 'physiological changes' that 'for years' were linked to ageing were now 'more appropriately attributed to disease, lifestyle or both' (WHO, 1993: 2). They further argued that,

> Because of the difficulty of distinguishing the true biological ageing process from the consequences of long-term lifestyle and working conditions, there are those who have advocated paying special attention to and studying people who 'age successfully'. This is likely to be especially important for a better understanding of the relationship of the older worker and the workplace and for identifying those who will need for their work to be adapted as they age.
>
> (WHO, 1993: 3)

Aligning themselves with those such as the investigators of the Baltimore Longitudinal Study of Aging (Chapter 6), who argued that profiles of 'abstract measures' based on averages observed in the population were likely to mask differences between 'true biological ageing' and illness, Ilmarinen and his fellow working group members proposed that the solution to extending working life hinged on understanding ageing in people who were exceptionally healthy and active. The contrast with Welford's cautious recognition of 'exceptional people' could not be more striking. As I argued in Chapters 5 and 6, exceptionally healthy people represented models of 'pure ageing' who could serve as normative and epistemic referents – the 'index cases' – to differentiate between individuals. In this regard, within the field of ageing and work, the WHO group were suggesting a return to the Fatigue Laboratory's aim to use physiological models 'to make trustworthy quantitative distinctions and discriminations between individuals' (see above). In other words, the aim to 'identify those who will need for their work to be adapted as they age', required the development of reliable physiological parameters, which were to be generated through the study of model organisms.

Similarly to the Fatigue Laboratory's approach, the WHO group configured these parameters in relation to an optimal relation – the perfect fit – between task demands and the worker's response. In an optimal situation, work demands are met with a worker's capacity to function and recover on time (WHO, 1993: 30–32). This capacity to recover is itself seen as related to the worker's 'reserves', much in the same way that McFarland conceived of these as the worker's ability to respond to a given stress (see above). From this perspective, in advocating that health maintenance was the determining factor in sustaining those reserves the WHO Work and Aging group approach appears as the realisation of McFarland's vision of preventative medicine (McFarland, 1956) in the workplace. Like MacFarland, the WHO group proposed that measuring work capacity should rely on a physiological understanding of functional capacity as relating to the capacity to adapt to given stresses. Such knowledge should be able to guide the production of the fit between man and machine for particular individuals as they age, 'taking into account individual needs and ensuring flexibility in the workplace' (WHO, 1993: 30).

In mobilising the role of stress and adaptation for enacting health, the WHO group and, later on, the WAI claimed inspiration from Selye rather than MacFarland. However, as I argued above, Selye can be seen as the populariser of Henderson's Harvard colleague Cannon's work on homeostasis. From this perspective, his work on stress represents an extension of the problem of 'external adaptation' to the social and political realm of the 1950s and 1960s (Viner, 1999), much in the same way that Henderson's *The Order of Nature* or Cannon's *The Wisdom of the Body* addressed the perceived issues of social order in the inter-war period. For Ilmarinen and the WHO group, Selye represented a widely accepted referent, easily recognised by their audience of occupational health practitioners. For them, it was also important that Selye had specifically linked the general problem of stress to ageing, arguing that 'people possessed a finite quantity of "adaptation energy" which was gradually consumed by the "wear and tear of life", leading to physiological ageing and death' (Jackson, 2012: 14). Selye's solution to this problem, clearly inspired by Henderson's writings, was to devise ways to protect individuals' reserves – the stores of adaptation energy – by 'living wisely in accordance with natural laws'.

Drawing on this linkage between the maintenance of 'reserves' and the 'wisdom of the body', the WHO group strongly advocated the implementation of 'health promotion' programmes in the workplace. In this, they were further supported by 'well established data' on the relationship between ageing and chronic illness incidence (WHO, 1993: 32), and by the role of lifestyles on modulating the onset of those conditions. They thus recommended 'periodic health examinations for all workers aged 45 and older' to provide individualised assessments of 'work capacity', which should be the sole 'criterion for hiring and retaining employees' (WHO, 1993: 35). Employers were also advised to incorporate programmes of health promotion in the workplace to retain older workers. This should be combined with flexible placement of workers in work roles and retraining of workers in accordance to work capacity.

Work capacity was, the WHO group proposed, in this respect, fully aligned with the transformations of the workplace in most developed countries, where the 'trend towards multiple skills, competency-based learning and retraining promotes better occupational flexibility' (WHO, 1993: 32). In this post-Fordist personnel management context, work capacity was an indispensable tool to support workplace and employment decisions. These decisions, the WHO group recognised, were undermined by an antagonistic approach to labour relations, and best fitted to organisations that fully involved the employer, employee and occupational health expert in the decision-making process. In other words, work capacity assessments, such as the Work Capacity Index developed by Ilmarinen and discussed above, were best suited to organisations that promoted cooperative, or corporatist labour relations.

From this perspective, the WHO and the work capacity approach can be said to also recover the configuration of power and expertise promoted by Elton Mayo, drawing on his work at the Fatigue Laboratory (see above). For Mayo, as for other members of what came to be known as the Pareto Circle organised by Henderson (Cot, 2011), democratic and 'grass-roots' movements', such as unions, involvement in organisational decision-making was undermined by their 'failure to appreciate the social importance of knowledge and skill' (Mayo in Bruce and Nyland, 2011: 395). Viewing conflict and workplace troubles as a result of 'maladaptation', he thus espoused a theory of 'democracy in industry' where 'knowledgeable worker elites' would be able to negotiate workplace rules and placements with 'human relations' experts and employers. The internal equilibrium of the firm relied on this 'corporatist' organisation, depending on experts 'to transform the subjectivity of the worker from an obstacle to an ally in the quest for productivity and profit' (Deetz, 2003: 140). For the WHO group, the partnership between employers and trade unions was best represented in the Nordic Model of labour relations, particularly that practised in Finland, where those 'social partners' – or their expert representatives – negotiate the terms under which they regulate the workplace among themselves. As Ilmarinen suggested a few years later in relation to the WAI,

> The maintenance and promotion of work ability requires good cooperation between supervisors and employees. However, neither can ensure that work ability will not change; instead the responsibility is shared between the employer and the employee.
>
> (Ilmarinen, 2009)

Sharing responsibility across labour process positions constituted the normative, institutional context for work ability assessment and management. However, in turn, the work ability assessment, the WAI, also made possible, *performed* the set of institutional relations that it required. This was ensured by the very procedure through which the assessment was conducted – self-assessment. A 'technology of the self', the WAI represents thus what Henderson imagined as a 'trustworthy quantitative distinction and discrimination between individuals', as it translates the paradigmatic treadmill experiments and their expression of 'maladaptation' into a process of auto-evaluation. Importantly, it is this procedure of self-evaluation that qualifies or *equips* the worker to share responsibility for his or her withdrawal from or continuation in the labour force.

As a solution to the ageing economy, the concept of 'work ability' and the WAI is not simply a technical fix, but it brings to bear a version of European values in labour relations in the 'ageing society'. As an embodiment of active ageing, it enrols individuals in the management of their own life course, health and wellbeing. In addition, it also offers organisations and firms 'trustworthy distinctions between individuals', equipping them to make decisions of retirement and adjustment of skills with age. Its main success is that it is an adaptable, transportable and flexible version of 'functional age'. As such, it is a compromise between the pessimistic and optimistic versions of the ageing society, supporting pension system reform from the 'inside' by emphasising the 'positive point of view of health and mental resources' of older people (Tuomi *et al.*, 1997: 7). It is also a compromise between the ideals of efficiency and authenticity, as described in Chapter 5, supporting 'a balance between work and personal resources' (Ilmarinen, 2009). It is thus not surprising that experts and policy makers have had such high hopes for the WHO Work and Ageing group and the WAI approach.

Conclusion

In this chapter, I have focused on work and technology within the ageing society. My argument was that as a means to understand and manage vitality and productivity in later life healthy and active ageing policies have become entangled with physiological, functional enactments of health and capacity to work. I explored how this conjunction was assembled during the twentieth century by the work of industrial psychologists, gerontologists and ergonomists interested in the factors that underpin the biological and psychological self-regulating capacities of labour. Using the WAI as the key agencement in this process, I traced its conceptual, methodological, political and normative entanglements. This led me to explore the genesis and transformation of the concept of functional age, from its origins within investigations of 'external adaptation' in industrial organisations to its current incarnation as 'work ability'.

In so doing, I have also described how the concept of functional age was traversed by an internal tension between two approaches to the problem of ageing and work. On the one hand, there was the physiological approach favoured by Mayo and McFarland, which used model organisms – such as the Olympic athlete – to identify the fundamental relationships shaping work capacity and fatigue. They used the exercise physiology laboratory as the index setting to enact these relationships. On the other, there was the Cambridge approach aiming to identify the key elements of skill by extraction from representative samples of the population. They used simulations of ordinary work situations to generate that data, but were dependent on their previous expertise on light, skilled work. This was perhaps the crucial factor in decreasing the second approach's influence in contemporary workplaces. As the economy transitioned to post-industrial processes, the key elements approach was replaced by a renewed interest in the factors shaping the ability of workers to adapt to external stresses, now conceptualised as health.

Whereas, in previous chapters, I have suggested that uncertainties still pertain to the measurement and enactment of the relationship between ageing and health, in this chapter I have instead proposed that re-evaluations of older people's labour have become inextricably linked to arrangements in the production and monitoring of health. I have however also suggested that this conjunction hinges on the mobilisation of modelling approaches to health, rather than solely on the population-, risk factor-based arrangements that are usually seen to support contemporary public health and biomedicine. In this respect, my proposal, following the argument of the chapter, is that in instruments such as the WAI the risk factor approach is included within the more encompassing normative and epistemic scaffolds allowed by the model organism approach to health. These enable the stabilisation of an enactment of health against which variations and regressions to the mean can be plotted. This also enables the extension of the network of the workplace towards life style and leisure activities performed by workers, including them in a wider arrangement of biological and psychological self-maintenance.

Chapter 8

Caring for ageing

Introduction

In the last chapter, I have proposed that contemporary technological democracies have experienced a reinforcement of the link between health and productivity. Seen from the triangular relationship I suggested is at the heart of the 'ageing society', this means that functional enactments of health have been deployed to imagine a perfectly efficient fit between humans and machines. In this 'human factors' tradition of research, experimentation on model organisms, such as the athlete, enabled re-evaluations of older people's labour, based on their potential productivity. This agencement was, I suggested, partially stabilised by its emplacement within the policy need to extend and enhance the employment of older people. The consensus around the need to prolong working life is justified not only by the personal and economic benefits that come from making a better use of 'existing human capital', but also, as discussed in Chapter 1, because withdrawal from the labour force is associated with increased financial transfers across sections of the population to pay for retirement schemes and health and social care provisions.

Relatedly, I have also argued that the key matter of concern of the 'ageing society' is the relationship between ageing, health/illness and mortality, that is to say, the extent to which there is a flexible association between morbidity and longevity. Because of the weighty technoscientific and political uncertainties around this issue, there continues to be an important focus on understanding and managing functional decline and disability associated with age, with 'ageing well' and adjusting successfully to illness in later life. For example, the EU has, in the last decade, adopted a series of initiatives about 'ageing well in the community' that emphasise the need to support older people in 'enjoying a healthier and higher quality of daily life for longer, assisted by technology, while maintaining a high degree of independence and autonomy' (Eurostat, 2012: 135). In this, ageing well programmes attempt to align two different types of ageing policies: one associated with programmes and instruments of active ageing – a consistent focus of attention in this book – and the other normally associated with the concept of 'ageing-in-place'. The latter is broadly defined as a person's ability to stay in their own home and community as they age, with instruments aiming to monitor and manage the drivers that prevent or lead to the institutionalisation of older people. While there is significant consensus about the possibility of bringing these two policies into line with each other, there are also substantial tensions between them.

Ageing well policies are hinged on the mobilisation of 'assistive technology' at home to coordinate active ageing and ageing-in-place. However, as Aceros *et al.* (2015) have argued, it is exactly this deployment of technological devices in the home and community that makes visible the complex dynamics between ageing-in-place – usually enacted as the safety and security of older people – and the autonomy and independence present in active ageing practices. This chapter focuses on tracing how this dynamics was brought to bear within contemporary technological societies. It argues that this dynamic shifted out of the interplay between two configurations. On the one hand, between the late 1960s and early 1970s countries like Britain or the US saw the consolidation of a social and political critique of the caring relationship embedded within the welfare state, and a re-evaluation of the role of extended families and communities in the care of older people. This enabled an acknowledgement of the care recipient as an autonomous, rather than necessarily dependent, person. However, this line of collective investigation also made autonomy a mutable quality, reliant upon the environment in which older people live, an artificiality it shared with the type of subjective rationality that Foucault identified as distinctive to neoliberal dispositifs of power (Foucault, 2004: 223). Nowhere, however, is this entanglement of agentic qualities of older people more evident than in gerontology's own inquiry into the psycho-physical capacities of 'prosthetic environments' for older people. This is the second configuration of interest in the chapter.

As discussed in Chapter 3, this search for new tools and 'designs for living' 'for the purpose of ameliorating the effects of functional changes that accompany the aging process' (Czaja, 1990: xii) is closely related to the reflexive accounting for the knowledge uncertainties that belong to the 'ageing society'. I draw on the example of the development of the Instrumental Activities of Daily Living Scale (Lawton and Brody, 1969) – a widely used tool in the assessment and planning of older people's care – as a window into how environmental gerontology deployed those uncertainties. Tracing the assemblage around the tool, I argue that its analytical decomposition of daily life – cooking, housekeeping, laundry, etc. – transposes an experimental form of reasoning to the home, whereby familiar objects are recast as items of 'control' or 'conditioning'. Deploying the human factors approach – already explored in the previous chapter – to domestic environments transforms action-in-environment into sets of components or factors shaping 'behaviour', now understood as a response to 'operant conditioning'. Seeking to ensure the 'matching of people and environments' (Lawton, 1970: 234), now in the home rather than the workplace, this approach came to underpin much of the current political and economic investment in 'gerontotechnology', and the use of assistive technologies and other devices in the homes of older people (e.g. Rogers and Fisk, 2010).

In the third section of the chapter, I argue that assistive technology's reliance on formally decomposing older people's relationship with their 'environments' is critically challenged by an understanding of caring practices that emphasises tinkering (Mol, 2008) and the order of the 'familiar' (Thévenot, 2007), that to say, by how care is crafted in the way persons accommodate themselves to their surroundings through trial and error. Tracing how this understanding partially overlaps with the first configuration of practices analysed in the chapter, I suggest that it makes possible the exploration of the emergent properties of familiar action that are enacted in, with and through the 'caring environment' (Pols and Willems, 2011). Realising this, the chapter concludes, should be an object lesson on how to revalue later life in contemporary technological societies.

Questioning old age care

One of the key transformations of the social and cultural framing of older people in the last decades has been the increased recognition that 'old age' does not have to be an institutionalised experience. As suggested above, questioning the insulation and lack of autonomy that usually accompanied – and is still associated with – institutionalisation in care homes has come to be linked to the movement labelled ageing-in-place. Ageing-in-place aims to highlight the fact that the physical and social environment in which people live is crucial to how they grow old, their health and their quality of life. In academic literature, the concept is underpinned by the exploration of the myriad dimensions underpinning the experience of place, understood often as a social, cultural and affective construct that is more than the sum of its parts (e.g. Kearns and Andrews, 2004). In the policy domain, ageing-in-place represents a concern with supporting older people remaining at home for longer, through care provision, housing modification, etc. In this latter domain, home as a place of care is intended to provide continuity of environment, the maintenance of autonomy in the community and social inclusion. Where this is no longer possible, this continuity should be provided through a 'sheltered and supportive environment which is as close to their community as possible, in both the social and geographical sense' (OECD, 1994: 37).

Critical social scientists have suggested, however, that this emphasis on continuity between home and formal care, and the policy transition towards ageing-in-place has occurred in the same economic and political context that saw a withdrawal of the state from key social functions (e.g. Milligan, 2012: 3–5). Drawing on this wider context of economic and political transformation, social scientists have argued that ageing-in-place, with its emphasis on independence, relies on the transference of agency and responsibility from formal to informal care networks, embedding caring in the community and the social environment. This alignment between ageing-in-place and neoliberal policies is not, however, a simple matter of rearticulating the role of the state in the management of old age, but is also intimately connected to a recasting and revaluation of older people's competence and right to manage their own lives. It is worth detailing briefly what is involved in this apparent contradiction.

Social and political scientists usually identify the 1970s as a period of transition with regards to the function of the state. Transformations in the economic fabric of societies in the wake of the oil crises of 1973 and 1979, with recurrent periods of 'stagflation' in advanced economies, are generally seen as the context for a re-emergence of ideological questioning about the role of the state and the professions in the provision of social security. This shift marks the emergence of neoliberal policies across the globe, underpinned by the configuration of markets in previously socialised goods or services, and the supporting of individuals' entrepreneurship capacities. In the ageing domain, this transformation is most usually associated with the articulation of the 'three pillar' concept in old age pension thinking by the World Bank (1994), whereby pension contributions are seen as a combination of non-contributory, voluntary and regulated savings. In the care sector, these changes materialised in the consolidation of what has become known as the 'mixed economy of care', where public, voluntary and private providers complement the informal care provided by families and neighbours. Genealogically, these changes have been consistently linked to the establishment of conservative 'thought collectives', such as the Chicago School of Economics (Harvey, 2005). What is less often explored is how the questioning of the model of the welfare state, particularly in relation to old age, was also linked to a 'progressive' re-evaluation of the power of the service user.

As we have seen in Chapters 3 and 4, thrift and the ability to care for oneself has been the key normativity of 'insurance societies' and the correlated shaping of old age pension schemes. While there was always opposition to the implementation of social security for fear of its effects on the behaviour of welfare recipients, particular from the liberal right, an emerging concern about the welfare state's formalisation and bureaucratisation of forms of social solidarity began to take shape in the 1960s. Focused around the need to develop more responsive services and encourage 'self-help' or informal care, this process was reinforced with growing policy interest in funding research and initiatives of neighbourhood care or 'grass-roots helping' throughout the 1970s. This in turn led to the funding and publication of what is usually recognised as a landmark document in the articulation of the idea of the mixed economy of care, the Wolfenden Report on voluntary organisations (1978). The Wolfenden report is usually seen as presenting a challenge to the orthodoxy of British social policy theory, where a shared vision of a more equal, more just society, with 'better' social services financed through redistributive taxation had been established since the 1940s. But where some would suggest that this challenge was underpinned by attempts to re-establish the status quo that had been deposed with the emergence of the post-war welfare state, it also relied on a complex integration between progressive political ideals, policy frameworks and social science's conceptual apparatus within the same time period.

In the British context, this integration is perhaps best embodied in the community studies approach (Willmott and Young, 1957; Bell and Newby, 1972), wherein the most significant example for our purposes is the studies led by Peter Townsend on the elderly. As we saw in Chapter 5, drawing on the same methodology, concepts and fieldwork site as Young and Willmott, Townsend collected interviews with older people living in Bethnal Green to demonstrate that the extended family and its associated moral obligations had survived and transformed with the establishment of the welfare state. As a Fabian socialist, Townsend's aim was not to wholly critique the foundations of the welfare state, but instead to reform its atomistic assumptions about its citizens, so as to understand older people 'as members of families' (Townsend, 1957: 227). By contesting the individualist assumption of bureaucracies in 'insurance societies', and its reliance on indexing need on chronological age, Townsend was able to render older people as knowledgeable, competent persons, managing a richly textured social environment, and to contrast this with the approach taken by social workers and other professionals in assessing the 'needs' of older people. In this regard, Townsend's studies can be seen as one of the origins of the concept of ageing-in-place, couched as they were within social science's epistemic and normative network which enacted Bethnal Green's family and neighbourhood relations as intimately linked to the spatial and architectural organisations of London's East End. This articulation of the attachment of older people to place led to the claim, developed in subsequent studies, that usage of residential care and hospitals was significantly determined by the family structure of the older person rather than by his or her physical or psychological 'dependence'. Institutionalisation was, from the older people's point of view, a 'last refuge' (Townsend, 1962), and not necessarily desired or beneficial as a form of care in 'old age'.

Townsend's studies were of key importance in establishing what Philip Abrams later labelled 'the independence and legitimacy of the client's point of view' (Abrams and Bulmer, 1986). Abrams himself, in studies funded by the Rowntree Memorial Trust (Abrams and Bulmer, 1986: xi), was concerned with the role of community in mediating the relationship between formal and informal care. But where Townsend was confident of the capacity of the state to incorporate the 'client's point of view', Abrams was instead convinced that the relationship between the 'public world of bureaucrats' and the private world of care was one of incompatibility and friction. As Abrams put it,

> Formal care is provided within the ambit of bureaucratic structured agencies; it is a matter of tasks to be performed by specified persons whose work is to carry out those tasks, within an overt hierarchy or accountability. By contrast, informal care is rooted not in commitment to tasks but in attachment to persons; it is a property of relationships, not of jobs; its dispositional base is involvement with other people, not the conscientious performance of a role.
>
> (Abrams and Bulmer, 1986: 4)

Although his sceptic stance on the possibility of a collaborative relationship between different forms of care softened before he died in 1981, his work openly challenged the assumption of a well jointed continuum between formal and informal care which had been advocated since the Wolfenden Report (see above). In this, Abrams also pointed towards the political economy of care as unpaid labour of social reproduction, further theorised by feminist social scientists such as Finch or Ungerson (e.g. Finch and Groves, 1983), which became more acute in the assumptions underlying the health and social care reforms of the 1990s (Graham, 1991).

The link between ageing-in-place programmes and the destabilisation of the institutions of the 'insurance society' can thus be seen as resulting from the paradoxical alignment between the emergence of neoliberal policies and the consolidation of empowering, enabling approaches to social services. In this, social science's decomposition of the role of the state in modern societies, and its questioning of the power of formal rationality in simplifying and distorting the abilities and needs of its citizens, was most notably articulated in Habermas' (1984) theory of communicative action. In this theory, the formal powers of the state exert a form of colonisation of the life world, that is to say, upon the common values and beliefs that develop through face to face contacts over time in various social groups, from families to communities. Similarly, over the years, Townsend's view of the old age institutions such as residential care became increasingly negative, such that he, with Walker and others, came to think of them primarily as forms of regulating 'deviation from the central social values of self-help, domestic independence, personal thrift, willingness to work, productive effort and family care' (Townsend, 1981: 22). Indeed, it was one of the key messages of critical social gerontology that what were seen as the 'problems of old age' were largely constructed as a result of the capitalist form of social and economic organisation, and in particular the combination of professional control and the state's interest in regulating the value of labour, and the increasing commodification of later life (Estes, 1979; see Chapter 10).

Such a critique is also deployed in governmental analyses of the emergence of the liberal welfare state, the 'insurance society' and the neoliberal paradigm, analyses which this book drew upon to explore the formation of the 'ageing society'. The argument in Chapter 2 was that old age care institutions were a dispositif of power-knowledge which has been questioned by the constitution of neoliberal or 'governmentalised' ageing policy initiatives. The latter rely on forms of knowledge-mediated administration supported by the gathering, processing and provision of information to guide the conduct and choices of individuals and organisations. In aiming to 'activate' older people, such programmes and instruments aim to unfasten the entanglements that are seen to hamper agentic properties in individuals. Ageing-in-place policies see this as best achieved when older people are supported within the home and community. As I have argued above, however, the link between active ageing and the ability to stay in one's own home is far from necessary or obvious. It is, rather, conditionally obligatory, made possible by a specific conjunction of epistemic and normative conditions whereby the 'subjectifying' apparatus of active ageing programmes came into alignment with an articulation of the voice of the client within formal old age care. The latter drew on the network of physical, social and cultural elements as the normative, legitimising grounds for such a voice – the voice of the lifeworld. In so doing, it became entangled with the artificial quality of subjective rationality that Foucault identified as distinctive to neoliberal dispositifs of subjectification, understood as 'the strategic programming of individuals' activity' (Foucault, 2004: 223).

It is my suggestion that this entanglement explains the paradoxical alignment between neoliberal programmes and 'progressive' critiques of the welfare state, and further between active ageing and ageing-in-place policies. The interlinkage relied on a shared enactment of behaviour as artificially generated, as malleable and manipulable. Whereas in neoliberal models such artificiality is mostly articulated in relation to a conceptualisation of the ontological nature of market competition, it however relies on configuring individual conduct as an epistemic object in the behavioural sciences, and most particularly within experimental psychology. Equally, although seemingly detached from the analyses and decompositions of the welfare state produced by critical sociology, the latter's concern for preventing a mechanistic reduction of human experience is underpinned by a recognition of the fragile social and political ecology that makes dignity, freedom and autonomy possible. The transposition between these modes of reasoning relies thus on a complex trail of relationships that is best explored through empirical analysis of a widely used instrument in the care of older people, which will be the focus of the next section.

Conditioning 'old age'

In order to understand how neoliberal subjectifying dispositifs became aligned with the aim to maintain older people at home for longer, it is not enough to link this entanglement to processes of restructuration – marketisation – of public services. As I argued in the previous section, this alignment relied upon a complex articulation between two different ways of enacting the artificiality of individual conduct. The intertwining of these two modes is a process specific to the domain of ageing, and how its assemblage shifted in the years heading to the foundation of the 'ageing society', particular in the US. In these years, debate and controversy about how to manage the commitments embedded in Medicare created the need for new tools and instruments to measure the health and functionality of older people (see Chapters 5, 6 and 7). One of the most important instruments created in this context was the Instrumental Activities of Daily Living Scale (IADL).

Widely used both in research studies and in clinical and community work practices, IADL assessment is a simple structured questionnaire that evaluates a person's capacity to perform household maintenance tasks, such as shopping, preparing food, taking medication or handling finances. Focusing on each of the tasks at a time, responses to each of the eight items in the scale will vary along a range of levels of competence, from independence in performing the activity to not performing it at all. The higher the score, the greater the person's abilities. Known for being easy to administer and flexible in its scoring and interpretation, IADL assessment is also preferred to other assessments of activities of daily living, such as Katz's (Katz *et al.*, 1963), because it is believed to measure functions underpinning autonomy and independent living at home. This focus on autonomy and the household is not a coincidence, and is key to the entanglement between diverging modes of enacting artificiality.

The IADL was devised by M. Powell Lawton and Elaine Brody in the context of the Philadelphia Geriatric Centre's expanding services following the establishment of Medicare. Working as a team composed of a social psychologist – Lawton – and a practitioner social worker – Brody – the tool was, from the outset, designed to serve providers' decision-making, service planning and management, linking such processes to procedures of geriatric assessment. In particular, given the complexity of evaluation methods regarding older people, the tool aimed to 'bring order to the planning process' linking different professional perspectives and fastening individual treatment trajectories to the organisation of services (Lawton and Brody, 1969: 183). Such a mediating role could only be achieved, Lawton and Brody argued, by substituting assessments derived from particular forms of clinical expertise with a focus on *function*. In so doing, the IADL brought to bear a critique of geriatric medicine, and its focus on the 'needs' and deficits of old age. Instead, Lawton and Brody suggested that, for the purposes of deciding or planning transference of persons between forms of care, the emphasis should be put on the individual's *competence* rather than his or her incapacities or needs.

Herein, in this focus on competence lies the distinctive approach proposed by the IADL assessment tool: that instead of focusing on needs, it makes abilities visible. The highlighting of positive function can, of course, be itself contextualised in the wider epistemic and political concern with the measurement of functional age, discussed in Chapter 7. There is, however, as I argued above, a specific embedding within the politics of Medicare, on the one hand, and the construction of an applied psychology in care settings, on the other, that is worth exploring in depth. First, the relation with the establishment of Medicare.

The aftermath of World War II saw the formation of a progressive movement that wanted to expand and organise the provision of welfare. This was nowhere more acute than in President Truman's 1948 proposal to create a system of universal health insurance coverage, in the spirit of the creation of the British NHS in that same year. Such proposals, as documented by Oberlander (2003), were met with opposition from the American Medical Association which, in coalition with the Republican Party, was publicly critical of 'socialised medicine' as it undermined the value of 'choice' and freedom. Not wanting to strategically link this issue to the wider problem of Cold War politics, particularly after the Korean War (1950–1953), key supporters of Truman's proposal, such as the unions, moved the focus of federally backed health care towards the 'elderly'. This was not only politically expedient but also meant that programmes could be supported by social insurance payments accrued since the 1935 Act. This proved a decisive move because, as Marmor put it, while 'it was one thing to write off socialism [,] writing off the aged would give the wise politician second thoughts' (Marmor, 1973: 28).

The growing political consensus around a health care programme for the 'elderly' created however a variety of uncertainties: how would these programmes be organised?; what would need to be built and arranged to serve the needs of this population?; what forms of practical expertise and knowledge would be required to address these challenges? As Achenbaum (1995) argued, such uncertainties formed the basis of a growth in the social and psychological research on ageing during the 1950s and 1960s, as researchers attempted to conceptualise the behaviour of older people and develop tools to assess their needs. Indeed, Lawton and Brody themselves recognised that because 'the use of formal devices for assessing function is becoming standard in agencies serving the elderly [...] a large assortment of rating scales, checklists, and other techniques in use in applied settings' (Lawton and Brody, 1969: 179) had been assembled in the previous decade. 'Rating scales, checklists and techniques' promised to ensure the values of bureaucratic efficiency through 'objective', scientific forms of sorting individuals' position within, and trajectories across, services, aiming to replace the formal rationality of age grading assessment of need with a substantive focus (see Chapter 5). This was particularly significant as older individuals were seen as being less able to make informed judgements about their own health and needs.

Securing the objective nature of function assessment, however, required the articulation of the IADL with a theory of behaviour, thus linking functional assessment to the mode of reasoning deployed in social psychology in the US in the decades after the war. Departing from the view that human behaviour is structured by degrees of complexity – themselves sedimented on hierarchised levels of what they labelled 'neuropsychological organization' – Lawton and Brody argued that 'physical self-maintenance' could be assessed by a simple scale focusing on key housekeeping activities (Lawton and Brody, 1969: 179–180). This argument was supported by Lawton's own work on what has become known as the environmental docility hypothesis. This hypothesis postulated an inverse relationship between function and the environment's power to shape behaviour, such that the physical and social conditioners of behaviour increase their effect as the functional status of the older person decreases (Lawton and Simon, 1968).

This hypothesis can, to a large extent, be seen as a deferential application of Kurt Lewin's behaviour function – $B = f(P, E)$ – which suggested behaviour was the result of combined individual and environmental factors. Seen as one of the founding ideas of the field of social psychology, Lewin's function is indicative of a form of reasoning rather than representing an actual equation. This approach, which Lewin called 'topological' or field theory, became the foundation of Lewin's own programme of experimental social psychology at MIT, in collaboration with Leon Festinger among others, from 1945 onwards (Patnoe, 2013). Having been introduced to Lewin's approach by his tutor at Bryn Mawr College at the beginning of the 1950s, Lawton pursued his interest in gestalt psychology by combining both research and clinical practice in his exploration of the role of 'stimulus structure' to diagnose underlying individual 'psychopathology'. Drawing on this experience of analytically decomposing the 'forces' shaping a behaviour, Lawton was quick to translate this way of reasoning in his role as a research psychologist in a geriatric centre moving towards the provision of semi-independent living for older people in the 1960s.

In reality, Lawton's involvement with the Philadelphia Geriatric Centre provided him with quasi-laboratorial conditions, as he was able to observe older residents going about their daily life in an institutional setting and, as a resident researcher, was free to trial instruments that could accurately differentiate their behavioural capacities. Furthermore, the Philadelphia Centre was ahead of the curve in terms of introducing semi-independent housing for its patients, so that when, after 1965, new federal programmes for purposely designed housing for older people began being implemented, Lawton had already accrued some experience in the domain of 'ageing and environment'. Lawton's point of departure was that as people age they increasingly become attached to – entangled in – the place where they live, but concurrently become more sensitive and vulnerable to their social and physical environment. This stepping stone had been established by a combination of his observations of elderly residents at the Philadelphia Geriatric Centre, and shared beliefs about the physiological and psychological implications of increased age (see Chapter 5).

However, rather than focusing on the dynamic of psychological traits alone, Lawton posited a 'theoretical mean adaptation level for individuals of different competence interacting with their environments' (Nahemow and Lawton, 1973: 27). Redolent of Henderson's 'external adaptation' concept (Chapter 7), but instead drawing on the idea of homeostasis, the adaptation level denoted a capacity to maintain function and enact a perfect 'match' between goals and environment. This – the environment press model – also proposed that at the margins of the adaptation level lay a zone of 'optimisation' where environments could be perceived and act as 'positive stimuli' on a person's behaviour. This is where physical and social conditions could be understood as 'prosthetic environments for older people' (Lawton, 1970).

The idea of a 'prosthetic environment for older people', as Lawton (1983) himself later recognised, had been first explicitly proposed by Ogden Lindsley. Lindsley, often credited with coining the term 'behavioural therapy', was a disciple of B. F. Skinner, becoming the Director of the Behaviour Research Laboratory at Harvard Medical School in the 1950s. In this context, Lindsley deployed Skinner's radical behaviourism against the recalcitrant problem of 'schizophrenia'. As

Rutherford (2009) has argued, the key to Skinner's approach to the mind was that it was configured as a machine, viewing behaviours as malleable, technological mechanisms. Working closely with Skinner, Lindsley constructed an experimental device – the operant box – which he deployed in rats, dogs and, subsequently, psychotic patients. The main outcome of this work was the stabilisation of the behaviouralist's experimental system as a 'method for discoveries of operant conditioning' (Lindsley, 1964: 418). This, in turn, reinforced Lindsley's view of psychologists as 'behavioural engineers', whose expertise was relevant in a variety of social problems.

Turning briefly to the problem of 'old age' after his own mother's death, Lindsley proposed a programme of experimental research of 'free-operant conditioning' to offset the mental and physical decline associated with age (Lindsley, 1964). His belief was that much of what he called 'retarded behaviour' observed in old age was in fact being reinforced by the environment where older people lived. He advocated that precise measurement of older people's 'behavioural potential' should enable the identification of behavioural mechanisms, leading, in turn, to the outlining of individualised environmental conditioning programmes. However, he did not see these as 'therapeutic' because they did not alter the 'behavioural deficits' themselves, or retard the ageing process; they were prosthetic in that 'they must operate continually in order to decrease the debilitation resulting from the behavioural deficit' (Lindsley, 1964: 422). Drawing on the analogy with spectacles and hearing aids, Lindsley identified a variety of possible prosthetic conditioning devices acting as stimulus, response or reinforcement agents.

Much of what Lindsley saw as deficits relied on a comparison with a 'normal adult', and a somewhat unsophisticated view of 'old people' as a homogeneous behavioural group. This was, in large part, a function of his condition of outsider to the field of gerontology. For most psychologists of ageing, Lindsley's attention to 'behavioural deficits of the aged' would have easily smacked of the deficit model of ageing they were trying to destabilise. Nevertheless, his idea of the prosthetic environment as a compensatory mechanism for loss of function in old age appealed to policy makers and planners embedded in implementing Medicare. From this perspective, Lawton's own proposal can be seen as fundamentally aligned with Lindsley's behavioural engineering approach. However, instead of looking at environmental prosthesis as reparation of loss of function, Lawton viewed the design of care settings as enabling and supporting 'homeostatic behaviour', or as Lindsley would have it, as therapeutic. This was because he viewed such settings as the enduring context for naturally occurring behaviour rather than as a modified form of experimental conditioning.

For Lawton, a prosthetic environment should work so as to maintain individuals within their 'optimisation level'. This level was not a generalised standard of behavioural capacity, but was seen to vary between individuals, according to their 'competence', which could be measured by ADLs. But maintaining this 'optimisation level' was a continuous and precarious dynamic:

An eighty-year-old who is showing signs of declining competence, but who is still able to function in accustomed surroundings is relocated. The environmental press increases beyond her tolerance level and her functioning is markedly impaired. A danger for the person of declining competence is that of being removed to a too-supportive environment. The person of moderate competence who, with some help, could continue to function in the community, might also adapt downward to the limited environment of an institution, that is, become too content with functioning in the zone of comfort. In this case, she finds herself in a situation where it is difficult to seek or to find stimulation. However, the environmental press is not so constantly low that she is really driven to the extremity of 'maldaptive behaviour'. She seemingly adjusts very well to the home environment but her powers decline markedly. What has happened is that in a situation with constantly low environmental press, the transactional balance adjusts with an adaptation level at a lowered press, which leads to diminution of competence.

(Nahemow and Lawton, 1973: 29)

In this powerful illustration, Lawton and his colleague Nahemow describe a situation where relocation from the 'community' to a more assistive environment is based on an erroneous assessment of competence. This leads to the person becoming under-stimulated, which might result in 'maldaptive behaviour'. If in the illustration, the maladaptive outcome is avoided, it none the less has enduring consequences for her level of functioning. From this perspective, Lawton and Nahemow use this illustration as the reverse of the prosthetic environment, its antithesis. Due to what they call the 'transactional balance' between person and environment, the person's surroundings pull her aptitude downwards. A prosthetic environment, by contrast, should push the person just beyond her 'adaptation level', enabling activities that were not possible in her previous situation.

Drawing on this, Lawton (1983) later extended the experimental enactment of behaviour to a wider understanding of psychological wellbeing and quality of life. Proposing a 'model of the good life', Lawton suggested that the 'objective environment' lay at the heart of the processes leading to life satisfaction. His proposal was that these were not mechanistic, universal relationships, there being wide individual variations in how environment affected dimensions of the 'good life'. In this, Lawton's proposal was redolent of McFarland's deployment of the notion of internal equilibrium to find an individual fit between man and machine, discussed in the previous chapter. Here again, the experimental, model organism approach to ageing was mobilised not as a knowledge procedure but as a foundation for a way of thinking about the environment and the organism. As such, it established a template for enacting this relationship as an individual, optimal, adaptive equilibrium. In an optimal fit between the home environments and the older person, however, the aim is the production of 'quality of life' understood as a differential between expectation and outputs of competence. From this perspective, the IADL enacts a functional version of health, where quality of life becomes subsumed in 'objective assessment' of function. In other words, it embodies expectations about 'desirable behavioural or psychological outcomes' (Lawton, 1983: 353), those being linked with autonomy and self-care.

The idea of 'prosthetic environments for older people' embodied a promise of an old age that did not necessarily equate with illness and low quality of life. This promise resonated louder when the components of the 'ageing society' began to be articulated in the 1980s, as analysed in Chapter 3. As Congressional committees, the Department of Housing and Urban Development, the Administration on Aging and the White House became increasingly interested in the health-related issues of housing, and the design of institutions such as nursing homes (Lawton, 1990; Czaja, 1990; Chapter 3), Lawton's approach became increasingly recognised as crucial to the solution of the technological problem of the ageing society. In particular, it presented an assemblage of instruments, concepts and practices through which it was possible to regulate the usage of care facilities and services. As policy makers dealt with the consequences of increased life expectancy on social care budgets, environmental gerontology offered the means through which functional maintenance could be made visible and accountable. Further, it offered a justification for such policies from the perspective of the user, as discussed in the previous section, as attachment to place and community became publicly recognised as a key dimension of quality of life and wellbeing of older people. In this regard, this research was already entangled with the policy aim of managing increased demand for care technologies and services for older people.

Environmental gerontology embodied not solely a vision for sorting people and matching them with suitable existing services and facilities but also expectations about the role of technology extending activity and health for older people. For Lawton and early environmental gerontologists, this meant principally housing design that was deemed suitable for different levels of 'competence'. Towards the end of the 1980s, within a new horizon of use of technologies at home (e.g. microwave ovens), a platform of collaboration developed between gerontologists and human factors engineers. This was not new, having been developed in relation to work environments, as described in Chapter 7. In fact, what appears to have happened is that, as the demands for industrial gerontologists waned, some of that expertise was transferred to the endeavour of developing assistive technologies for older people. The uncertainties involved in this transposition were acknowledged. For example, in a section on 'home activities', most probably written by Lawton himself, for the report on *Human Factors Research Needs for an Aging Population*, already referred to in Chapter 3, it was argued that,

> There has been tremendous growth in the number of in-home services [...] and in the development of assistive technologies that can extend the ability

of the elderly to live independently. However, before the benefits of these technologies are fully realized, several objectives must be accomplished [, such as] the development of methods to match services and technologies with individuals. A valid functional assessment protocol is important to achieving the latter goal. In addition, efforts must be directed toward designing products and devices that are accepted by the user population. [...] Clearly, this is an area where human factors researchers can make contributions. [However] it remains unclear to what extent the applications of technology to the physical environments or technological aids within the home mitigate the need for supportive assistance.

(Committee on an Aging Society (1988: 21–22)

The author's ambivalence towards the uncertainties surrounding the use of assistive technologies 'to extend the ability of the elderly to live independently' is clear. There was concern about the ability to match technology with individual needs, which justified the continued use of assessment tools such as the IADL. There were also concerns relating to the acceptability and usability of technologies, which needed the contribution of human factor engineers and designers, and presumably of social scientists, to investigate and assess public attitudes towards these technologies. But, most importantly, it was not known whether assistive technologies could replace care workers in providing 'supportive assistance'. This pointed towards the lack of research on 'the appropriateness and effectiveness of various assistive technologies' (Czaja, 1990: 22). In other words, it was not clear if assistive technologies could ever come to constitute 'prosthetic environments for older people'.

The problem was that in mobilising the 'human factors' approach, environmental gerontology, and later gerontechnology, also threaded upon its focus on efficiency and productivity. Relying on Morehouse's (1959) definition of human factors' ultimate aim as being 'the optimal utilization of human and machine capabilities to achieve the highest degree of effectiveness of the total system', researchers focused increasingly on 'task analyses', decomposing domestic activities such as cooking or cleaning into systems (e.g. Clark *et al.*, 1990). Such decomposition was aimed at insulating particular functions of behaviour, so as to support their technological substitution with the declining abilities of the older person. In this regard, by drawing on this ideal of effectiveness, human factors researchers aligned their work on the idea of prosthesis as compensation proposed by Lindsley. As technology development was closely related to the restoration of a system of activity that had been disrupted by functional impairment of the human component, the idea of optimal adaptation was enacted as the restoration of equilibrium, rather than of creating a condition of dynamic equilibrium. This was recognised by one of the central figures in the field, James Fozard, who claimed that contrary to compensation, enhancement of function had 'virtually no attention paid [...] by human factors and ergonomics' researchers (Fozard *et al.*, 1993: 8).

One of the reasons for this return to prosthetics as restoration of function is that efficiency had come to subsume other values. This was, on the one hand, because of the way in which tools such as the IADL enacted a functionalist version of health as the capacity to adapt to the external, environmental 'press'. Designed to assist the provider in finding the optimal fit between persons and environments, functional assessment tools worked as mediators between persons and environments, enabling a translation, or equivalence, between them. Thus, despite Lawton and others' doubts about the capacity of assistive technologies in promoting 'desirable behavioural outcomes', it established a relationship between a stabilised set of 'tasks' or behaviours and an array of tools and mechanisms that could compensate for them. Strikingly, this collection of tasks, used to assess competence and autonomy, was mostly home based (excluding shopping). The efficiency of environments for older people became, in this process, equated with their capacity to enable autonomy within the home. As a result, valuations of efficiency of assistive technology, as I argued in Chapter 3, came to incorporate assessment of health and autonomy.

This relation of equivalence justifies the increasing identification of assistive technologies as an economic opportunity, rather than just a burden on social care budgets. In this forthcoming but 'rapidly growing silver market', technologically innovative products designed for older adults, particularly those with compromised physical, cognitive, or social functioning, are seen to be of central importance (Rogers and Fisk, 2010). Indeed, market growth hinges on such technoscientific promises, a vision where personalised care for older people is populated by technological devices:

> Robots accompany frail older adults while they stroll around the house or use the bathroom; personal computers provide cognitive or physical training programs; smart home environments support people with sensory, mobility, or cognitive decline; and robotic animals play a significant role in the social and emotional life of older people with dementia. Future cohorts of older adults will benefit from a full range of technology products designed to support them as they 'stay connected' and age well despite accumulated loss experiences.
>
> (Wahl *et al.*, 2012: 154)

Caring for ageing with technology

The exploration of the ways in which neoliberal subjectifying dispositifs became aligned with the aim to maintain older people at home for longer, led us to link this process with the development of an equivalence between autonomy and efficiency in assessments of the value of assistive technologies for older people. Enacted as prosthetic environments to produce 'desirable behaviour outcomes', assistive technologies came to embody a functionalist version of health, underpinned by an experimental deployment of 'optimal adaptation' between 'man and machine'. This, in turn, relied on a longstanding assemblage of tools, concepts and procedures which understood behaviour as a function of 'conditioning'. However, in this process, Lawton's own proposal of prosthetics as enhancement rather than compensation for lost function was weakened by the way in which tools such as the IADL configured autonomy and agency as domestic competencies. That is to say, in limiting the range of practices that count as 'activity', the IADL facilitated, I suggested, the establishment of a *domestic*, house-bound technological paradigm for assistive technologies for older people.

As Aceros *et al.* (2015) have indicated, the domestic confinement of assistive technologies and telecare also points towards the uneasy, contingent relationship between active ageing and ageing-in-place policies. The appeal of a human factors approach to technological development is thus immense because of how it promises to solve the tensions within this relationship. As in relation to health and activity, the technoscientific promises linked to prosthetic environments for older people are highly contentious and contested solutions to the 'ageing society'. While technology forecasters propose scenarios where robots and pervasive technologies support older people's activity and connectivity (above), others point to the contrast between projection and reality. For example, research drawing on data from the English Longitudinal Study of Ageing identified a very low uptake of assistive technologies, it being significantly correlated to levels of wealth of individual users (Ross and Lloyd, 2012). Further, the two main assistive devices used in the population over 50 were mobile personal alarms (2 per cent) and alerting devices fixed to the home (4 per cent). The disparity between the technological horizon of autonomy and the reality of surveillance of older people's behaviour through technological devices is concerning, and poses important questions about the role of technogovernance in later life (Tulle and Mooney, 2002; Mort *et al.*, 2013).

One key issue in this regard concerns the shaping of home life through the conflation of activity and domesticity, embedded in the human factors approach to assistive technologies. Because this approach carves out and makes explicit what normally belongs to the mundane realm of activity, the devices it proposes embody specifications on how to manage and organise the household. It is not only that this specification is very likely to be Western, middle-class and gendered – epitomising the 'optimal adaptation' between 'man and machine' – but also that it compels a shift between different ways of inhabiting the world. This difference is best expressed in Thévenot's (2007) concept of 'regimes of engagement'. In this conceptualisation, human action can be emplaced in three configurations: one that deploys coordination through public orders of worth, a second that organises action through its contextual adequacy and a third that is underpinned by routine, taken-for-granted ways of doing things. In the latter, which Thévenot labels 'regime of familiarity', adequacy of a particular action or attachment is grounded in how the body of the person in question is comfortable with the specific arrangement of things around him or her. In this regime, activity is distributed between human and non-human actors in the form of mutual attach-ment, making it difficult to distinguish who/what is acting at any given time.

This situation differs from 'planned action', where autonomy depends upon a functional grasp of objects, enabling individual strategic objectives to be delineated through the use of elements in the environment. Underpinned by the ordinary lan-guage of individual accountability, this regime is easily captured within pro-grammes and instruments that require individual responsibility or, as in the case of ageing-in-place policies, that enact an individualised – optimal, efficient – fit between individuals and their environments. Seen from this perspective, assistive technologies' main effect is to bring the 'regime of planned action' to bear in the household, closing down the possibilities for engagement with the world. Indeed, ethnographic research on telecare devices at home shows that it is left to users to open and maintain the pragmatic compromise between regimes of engagement (e.g. Aceros et al., 2015). This form of work is complex because it requires attempting, through trial and error, to integrate technological devices within a distributed array of components which are not easily detached from each other. The work of integra-tion carries the chance of problematising elements of the household that were hith-erto seamlessly combined. Because knowledge of familiar attachment is 'fragmentary and specific to a customised thing and [does] not identify standard objects in their entirety' (Thévenot, 2007: 416), mainly expressed in embodiment, it is difficult to test such arrangements. They require continuous tinkering.

A reading of Mol's (2008) work on chronic disease management models would position this tinkering within the realm of care. Using the case of diabetes management, Mol argues that the patient is enacted within two logics: choice and care. The logic of choice emphasises individual determination and control over diseases, the pursuit of aims and the deployment of strategy; the patient is understood as an autonomous individual, capable and willing to be involved in his/her own treatment through the medium of choices and decisions. For this purpose, a plethora of information and evidence is produced to support the patient in her choices, and work is developed to give the patient the cognitive means to interpret and pursue the information. Moral worth is attached to the patient who is informed and involved as health and social care systems increas-ingly rely on such practices of self-management.

The logic of care, on the other hand, values some degree of passivity, of dependence, of 'letting go'; it focuses on collective forms of managing the illness and on continuous adjustment to an evolving relationship between disease and people's lives. Information is still important but it is not seen as entailing particular forms of cognitive agency or moral engagement with illness. The value of decision is replaced by that of 'tinkering', a creative, collective engage-ment with illness that is focused on the mundane, the 'just this, now' aspects of disease management. Information is brought to the fore to address particular problems, without wanting to establish an overarching path. This distributes responsibility in time and across actors involved. It is a logic that is redolent of the 'familiar'.

This understanding of care challenges the assumptions embodied in the alignment between ageing-in-place and active ageing. To be precise, it proposes that this alignment, hinged as it is on the 'artificiality' of individual behaviour, is reliant upon maintaining a compromise with practices that enact older people as dependent, in the sense offered by Thévenot and Mol. This compromise in turn requires that these two regimes or logics are not included or colonised by one another in full. One can hypothesise that the successful introduction of assistive devices for older people is a process whereby there is choreography between passivity/dependence and autonomy/agency (Gomart and Hennion, 1999). On the other hand, however, the lack of recognition – the invisibility – of this work of making familiar plays a key role in sustaining the politics of technoscientific promises in the 'ageing society'. Further, it reduces the number of ways older people can engage with the matters of concern of the 'ageing society', by only including them when equipped with a plan and a will to autonomy.

In this, as I suggested in Chapter 2, the social sciences have some responsibility. As I discussed above, in wanting to mend and liberate older people from the institutions and environments that turned them into 'dependants', the social sciences had to investigate the factors that enhanced or prevented activity and autonomy. In this respect, I argued that there was continuity of approach between the functional enactment of autonomy in social psychology – and Townsend's earlier work – and critical social science's denunciation of the political and economic conditions that made older people 'passive' recipients of care – evident in Townsend's later work. In wanting, once more, to reinstate autonomy and agency in our investigations about ageing and assistive technologies, we might be feeding into the normative consensus of the 'ageing society', and its aims to enhance efficiency by supporting extended working lives and older people staying in their own homes for longer. Opening assistive technologies to the 'regime of familiarity' enables recognition of the pragmatic pluralism of the 'ageing society' within social science.

Whether this recognition can be extended to the design and implementation of assistive technologies for older people depends upon opening this domain to user participation and critical design approaches (Wilkie and Michael, 2009). This might entail abandoning the commitment to exclusively enhance autonomy and choice for older people, and exploring how assistive technologies might be entangled with different enactments of the 'good later life' (Moreira, 2010). What these might be is uncertain, as closure around autonomy and 'optimal' relations between older persons and environments has left uncharted other ways of being old that do not quite fit in the dependent–autonomous dichotomy. Further, due to the dynamic of the familiar and care, models of innovation that rely on a division between experts and investors, on the one hand, and users and publics, on the other, might not be fit for purpose. Instead, what might be required is a practice of distributed innovation (Wynne and Felt, 2007). Applied to the assistive technologies domain, this will have to combine experimentation with trial and error, and a form of user involvement that is open to using proximal forms of communication, on multiple channels and formats (Moreira, 2012b).

Conclusion

In this chapter I have explored how assistive, environmental or ambient technologies have become central to the 'ageing society'. I have argued that their significance derives from how they are seen to align ageing-in-place policies with active ageing instruments. I have suggested that this alignment was contingently enabled by a match between the artificiality of behaviour within social psychological models of ageing, and the artificiality of conduct in the neoliberal governing of the ageing society.

I have further argued that while this alignment produced a technoscientific vision of old age where robots and pervasive technologies enhance autonomy in the face of function loss, this vision is also sustained by making invisible the routine ways which integrate the more mundane technologies in current use into everyday life. This questions the sustainability of the regime of technoscientific promises upon which the ageing society is founded. In the next chapter, I will continue to question this regime by focusing on technological promises to extend longevity and 'health span'.

Chapter 9

Biomedicalising ageing?

Introduction

In the last two chapters I have explored and opened up two of the technoscientific promises articulated through the assembling of the 'ageing society': activity and technology. I have traced how these two epistemic domains have become entangled with functional enactments of health. This was reliant on the deployment of experimental forms of reasoning underpinned by an 'optimal adaptation' between individual organisms and environment that is only visible in the laboratory. Employing this 'model' approach enabled researchers and policy makers concerned by the challenges of the 'ageing society' to propose scenarios of extended working lives and ageing-in-place which attempt to bound the range of possible ways of being and becoming old. These are, in turn, as I have suggested a number of times in the book, conditionally entangled in the possibility of changing the relationship between mortality and morbidity or illness for individuals and populations, that is to say, in the prospect that it might be possible to technologically alter the ageing process itself.

The technoscientific promise of applying the techniques of biology to the problem of ageing is nowhere more visible than in the emergence of 'anti-ageing medicine'. Seeking 'the advancement of technology to detect, prevent, and treat aging related disease and to promote research into methods to retard and optimise the human aging process', the anti-ageing movement proposes that 'the disabilities associated with normal aging are caused by physiological dysfunctions which in many cases are ameliorable to medical treatment' (American Academy of Anti-Aging Medicine, 2008 in Fishman *et al.*, 2010: 198; see also Mykytyn, 2010). This idea that 'normal ageing' constitutes a target for medical intervention challenges some of the more established assumptions about the role and functioning of modern medicine, which is supposed to prevent or cure illness. But this, as many social scientists and philosophers have suggested, itself hinges on a problematic assumption: what is 'normal', in relation to what, and for whom?

Within medical sociology, an important tradition of thought has conceptualised medicine as a 'moral enterprise', whereby social deviance is understood within a medical framework and managed through interventions on the body of the 'patient' (e.g. Conrad, 2007). From this perspective, medicine relies on pre-existing, mundane categorisations of persons as morally deviant to generate and legitimise definitions of disease and boundaries of knowledge and treatment. Extending this notion, social scientists have argued that medicine has, in the past decades, widened its jurisdiction and gradually focused its attention on 'normal behaviour', as it is seen to contain the seeds for the development of pathology. These changes have become crystallised in the concept of biomedicalisation (Clarke *et al.*, 2003).

Another perspective takes as its point of departure modern medicine's aim to understand pathology in terms of its relation to normal, physiological function. Famously, Canguilhem (1978) suggested that the nineteenth century laboratory approaches to pathology had eroded the ontological distinction between health and disease, illness becoming increasingly defined as a measurable deviation from a statistical norm (see Chapter 6). With the consolidation of molecular biology in the post-war years, scholars have argued that medicine has moved further away from clinical characterisations of illness towards using the experimental biology laboratory as a tool to understand and manipulate causes of disease, by redefining clinical cases as phenotypical expressions of interactions between genes and the environment. However, as Keating and Cambrosio (2003) have argued such uni-directional understanding of the relationship between the laboratory and the clinic ignores the myriad ways in which biomedicine relies on the constant interaction and intermingling between definitions of normal and pathological. Rather than a systematic application of biological standards and products to clinical work, bio-medicine is structured, they argue, around hybrid *bioclinical entities* that enable the difficult coordination between bench and bedside.

In this chapter, I propose that an analysis of the 'biomedicalisation of ageing' requires insights from both perspectives. On the one hand, there is a need to acknowledge and understand current scientific and commercial investment in age-retarding interventions and the molecular processes underpinning them. While these can be seen as associated with the reorganisation of health care around technological intervention and the modes of prevention and consump-tion, it is also important to realise that these processes, while significant, remain at the margins of mainstream biomedicine. As I will show in this chapter, there are both institutional and epistemic reasons for this.

One of the distinctive claims of the biology of ageing is that it provides the basis for an alternative model of understanding and managing diseases that are commonly associated with age, such as heart disease, stroke and cancer (Butler *et al.*, 2008). Rather than pursuing the disease-specific model that has been para-mount to the historical consolidation of biomedicine, biogerontologists argue that increases in health and 'health expectancy' are more likely to be achieved by focusing research on the common biological basis of all age-associated dis-eases. In reconstructing these diseases as part of a wider set of disorders linked to cellular senescence (e.g. Campisi and di Fagagna, 2007), biogerontology can be seen to be discursively participating in the aim of reducing pathology to biology through a 'molecularisation of ageing'. However, without consistent recourse to a body of established clinical practice and clinical research on ageing, biogerontology has struggled to stabilise the hybrid, collaborative platform of knowledge production that is characteristic of biomedicine. This is the result of both a lack of recognition of biology of ageing by the biomedical establishment and a rejection of the status quo by biogerontologists.

The chapter first discusses the conceptual and empirical basis underpinning the 'biomedicalisation of ageing' thesis. I argue that the technological promises of ageing research contrast with an everyday reality dominated by what is best described as a mundane focus on health management practices. While these are entangled with neoliberal forms of governmentality in which the production of optimal forms of health ensures individuals' civil and political rights, they are nonetheless limited by the unstable sociotechnical networks in which they circulate. The chapter next moves to exploring the reasons behind this instability, by following the debates and controversies about the shape of knowledge production institutions concerned with the biology of ageing in the US, France and UK. I show that only in the US, through a strategic focus on research on Alzheimer's disease from the 1970s onwards, was it possible to create solid lines of communication between policy objectives, commercial investment, research and clinical practice. This model, underpinned by the creation of a bioclinical entity, has been the object of a renewed critique by biogerontologists, furthering uncertainty about how to bring to bear a 'new biology of ageing' (Partridge, 2010).

Biomedicalisation of ageing

The origins of the expression 'biomedicalisation of ageing' can be traced to the publication of a paper in the late 1980s, where Estes and Binney proposed that there was a growing encroachment of the 'biomedical model' in the understanding and management of ageing in the US (Estes and Binney, 1989). This way of thinking that 'focuses on individual organic pathology, physiological aetiologies, and biomedical interventions in aging', they argued, was spreading through scientific practices, professional training, policy initiatives and lay public engagement. Key in this process was the consolidation of a reductionist, individualising approach to ageing that relied on the politically endorsed close association between basic sciences and disease oriented clinicians, a configuration particularly evident in the contemporaneous interest in Alzheimer's disease. This disease orientation was, in their view, fuelling a renewed interest in ageing among leading clinicians and policy makers and contributing to the 'growth of a multibillion-dollar medical-industrial complex' (Estes and Binney, 1989: 594). This they saw as part of the dynamic of the biomedicalisation of ageing which would eventually see the emergence of 'sources of resistance' (see also Zola, 1988).

Estes and Binney's paper is better understood if we contextualise it both historically and intellectually. In the 1980s, as we explored before, there was an emerging concern with the consequences of the 'social security crisis'. The National Institute of Aging, created in 1974 with a broad agenda of research on ageing, had seized on this public issue and progressively moved towards a focus on diseases of old age, particularly Alzheimer's disease (see Chapter 6). In parallel, since the 1970s, there had been a consolidation of a critical approach within medical sociology (Zola, 1972) and within social gerontology (Estes, 1979), reacting against what they saw as a functionalist validation of medicine. Instead, they argued that professionalisation of medicine and social services transferred autonomy and means of control from the client to experts, enhancing and legitimising power asymmetries in industrial societies. It was argued that these processes had been compounded by a bureaucratisation of health care and an increasing 'technologization' of medicine (Mechanic, 1977).

In this context, Estes and Binney's paper was aimed at setting a research agenda for the interface between medical sociology and social gerontology, by enabling a focused attention on the processes by which policies, institutions and practices increasingly managed older people through medical assessments and interventions (e.g. Bond, 1992; Kaufman, 1994). Its relevance was enhanced as social scientists documented shifting clinical thresholds for treatment and increased use of health technologies, and the associated rise in spending on health care programmes and services (Moreira, 2012a). In particular, the embedding of technologies in the diagnosis and management of illness had led to the suggestion that technological possibility rather than clinical reasoning was driving medical practice, in what Koenig (1988) labelled the technological imperative of biomedicine. Drawing on this, Kaufman and colleagues argued that this process had reshaped the normative grounds for clinical interventions in later life, leading to 'the transformation of the technological imperative into a moral imperative; and [...] the coupling of hope with the normalization and routinization of life-extending interventions' (Kaufman et al., 2004: 731; also Joyce and Loe, 2010).

By suggesting that there were significant shifts in the drivers of the process of biomedicalisation of ageing, Kaufman et al. were also drawing on a rearticulation of the concept of biomedicalisation itself proposed by Adele Clarke and colleagues (2003). In this, Clarke and colleagues had proposed that Conrad's concept of medicalisation was no longer wholly adequate to understand contemporary medicine exactly because it ignored the dynamic role played by knowledge, technological innovation and commodification in health care. Instead, they contended that,

> [b]iomedicalization is reciprocally constituted and manifest through five major interactive processes: (1) the politico-economic constitution of the Biomedical TechnoService Complex, Inc.; (2) the focus on health itself and elaboration of risk and surveillance biomedicines; (3) the increasingly technoscientific nature of the practices and innovations of biomedicine; (4) transformations of biomedical knowledge production, information

management, distribution and consumption; and (5) transformations of bodies to include new properties and the production of new individual and collective technoscientific identities.

(Clarke *et al.*, 2003: 163)

Where Estes and Binney had seen the growth of the medico-industrial complex (see above) as a result of biomedicalisation, Clarke and colleagues perceived commercialisation and commodification as permeating the technologies, institutions and knowledge produced and used in care settings. This knowledge and these technologies rely on the monitoring and management of health risk, which in turn are underpinned by an infrastructure of data collection and the creating of markets of biomedical information (genetic databases, clinical registries, 'quantified self' applications, etc.). Such technical assemblages are in close interaction with new forms of bodily engagement whereby individuals are empowered to know their genetic and biomolecular make-up, produce and maintain their own health and expect the tailoring of treatment to their 'bioclinical profile'. Biomedicalisation of ageing is one of the key areas where the general processes of biomedicalisation are seen more acutely (Fishman *et al.*, 2010; Joyce and Mamo, 2006; Joyce and Loe, 2010).

One significant example of the 'reciprocally constituted' process of medicalisation in the ageing domain concerns the commercial release of products aimed at 'telomerase activation'. Telomerase is the enzymatic process by which telomeres – structures on the end of chromosomes, which regulate cell replication – maintain their structure and function. Although its theoretical existence had been predicted since the 1960s, it is generally recognised that telomerase was first described by Carol W. Greider and Elizabeth Blackburn in the model organism *Tetrahymena*, for which they received the Nobel Prize for Medicine in 2009. Initially developed within the context of basic research on cell replication, the potential biomedical applications of the science – and in particular its relevance to understanding dysregulated cell replication in cancer – became quickly pressing with Greider's transfer to Cold Spring Harbour laboratory. This laboratory had, under James Watson's direction, flourished on the promises of genetics and molecular biology for medicine and health, and saw Greider's earlier work as part of its medium-term plans.

Its investment was such that by 1990, drawing on the development of a way to probe telomere length, Greider and colleagues had established a firm connection between telomerase and cell senescence, while others had confidently established its links to cancer (Blackburn *et al.*, 2006). Such discoveries underpinned the foundation of the Geron Corporation in 1992, a biotech company which would be described soon afterwards as 'the first biopharmaceutical company to focus exclusively on the development of therapeutic (and diagnostic) products for age-related diseases based on new molecular insights into the molecular and cellular mechanisms of aging.' (www.bio.net/bionet/mm/ageing/1996-May/002331.html). Counting Greider, Watson, Robert N. Butler (as former Director of the Institute of Aging) and Leonard Hayflick in the Scientific Advisory Board and backed by wealthy investors, Geron was tipped to become a key player in the bulging bio-technology field of the 1990s.

However, the Geron Corporation quickly became embroiled in complex legal disputes about their techniques of handling stem cells, being identified as the focus of much of the Republican-led political attack on such technologies in the early 2000s. Deciding to leave politically sensitive areas of research to focus on disease-specific programmes in 2002, Geron granted TA Sciences the licence to develop and sell a product they had been working on as a possible telomerase activator. Derived from a herb used in Chinese medicine – *Astragalus membranaceus* – but known commercially as TA65, Geron claimed it enhanced telomerase activity by protecting chromosomes and preventing the cells from becoming senescent. This claim was, however, unsupported by clinical studies or consistent results in the lab. Further, as a newcomer to the biotechnology industry, TA Sciences' CEO, Noel Patton, needed time to build networks in research and technology assessment and it was only in 2009, when the Nobel Prize was attributed for the discovery of telom-erase, that he was able to consider expanding beyond the few anti-ageing clinics that earlier had been prescribing TA65 to their private clients. Crucially, this expansion required linking with former Greider collaborator and Geron executive Calvin Harley, as well as with Blackburn and her student Maria Blasco.

With scientific backing from Blackburn, Harley had by now founded a new company – Telome Health – focused on testing telomere length, and Blasco was soon to commercialise her telomere measuring technique in Europe through Life Length. This loose network of biotech companies emerged into public view during 2011 when Blackburn announced that Telome Health was going to start providing telomere length testing as a gauge of individuals' 'underlying health' and risk of developing 'diseases of aging' (Blackburn in Dickinson, 2011). A controversy was subsequently sparked when journalists, seeking the views of Greider, her fellow Nobel Prize winner, learned that she, now a professor of molecular biology at Johns Hopkins University, questioned the evidence-base of her former colleague's claims, arguing that 'the science really isn't there to tell us what the consequences are of your telomere length' (Greider in Pollack, 2011).

Their disagreement fuelled public debate about the commercialisation of biomolecular testing, the possible uses of this type of information by insurance companies and the legal frameworks that regulate the release and access to such tests. This was further compounded when, in 2012, a class-action lawsuit was filed accusing TA Sciences of deception for advertising TA65 as capable of reversing the effects of ageing (Borrell, 2012). In a letter to the Federal Trade Commission of July 2013, the director of the Consumer Protection Coalition pointed out that animal tests of TA65 had not shown substantial age-retarding effects but that it had been to linked cell proliferation and cancer, so that 'at a staggering cost of $2,200 for a three-month supply, American consumers face substantial harm if TA proceeds to market without substantial scientific evidence'.

While the links between TA65 and Telome Health would appear to provide some scientific robustness to the product, it was revealed that neither was the test seen as valid nor the product as effective. Instead, both companies saw the consumer as being the driver of clinical uptake, and relied on internet sales for their business models. In this, however, they were both aware of the dangers of having their product's scientific credentials discredited by relying on networks of circulation outside the clinic. To offset this possibility, Telome Health likened telomere length measuring to currently available routine health assessment techniques, and emphasised that tests were provided through qualified physicians 'who understand and interpret the results for their patients' (Matlin in Connor, 2012). Further, champions argued that most of the interventions thought to affect telomere length were within the domain of health maintenance rather than through the application of medical technologies. This meant that TA65, being presented to regulatory agencies as a 'novel food', could escape the level of scrutiny of evidence that is usually applied to pharmaceuticals. But, and importantly, this strategy also revealed an insufficient interest in the product by clinical researchers who could deploy it in clinical trials.

This, in essence, is the paradox that permeates the biomedicalisation of the ageing process: the emphasis on health maintenance and the reliance on the consumer to commodify his/her own health is the flipside of a lack of penetration in the channels and networks of mainstream biomedicine. This is most evident in the recognisable 'outsider' status of anti-ageing doctors and scientists (see above) but derives from a longstanding epistemic and normative divergence between biological gerontologists and biomedicine that dates back to the middle of the last century. In the sections below, I trace the complex process through which this divergence came to be.

Diversity in ageing research institutions

As we saw in previous chapters, the mid twentieth century saw the emergence of the 'problems of ageing' in most industrialised societies. One of the key shifts in this was the increasing focus on measuring and understanding the drivers of efficiency of the existing 'stock of population', not only in terms of longevity but in relation to health, disease, etc. (Chapter 3). Such concerns, however, did not encounter a well formed and agreed set of epistemic and political commitments. Instead, there was a diversity of possible pathways being explored about how ageing should be emplaced and managed in knowledge making practices and institutions, which became the unsettling foundations of the 'ageing society' as we know it.

To a significant degree, this diversity of approaches was already evident in the formative years of the field of research on ageing. For example, as Park (2008) has documented, the seminal conference on the biology of ageing sponsored by the Josiah Macy Foundation in 1937 was ridden by disagreement between the various contributors, mostly around the 'parameters' and 'standards' by which to contrast normal and pathological ageing. While some, like Cowdry, proposed that quantitative differences in physiological functioning should replace simple, dichotomous views of normal and pathological ageing, clinically minded participants argued for a clear distinction that could be applied to individual cases. These differences were brought to bear a few years later in a key, but much-neglected, series of meetings sponsored by the CIBA Foundation between 1954 and 1956 aimed at 'the encouragement of laboratory and clinical investigations relative to the problems of ageing' (Wolstenholme and Cameron, 1955: I; see Chapter 5).

In its third meeting, focused on the 'Methodology of the Study of Ageing', discussion focused on the relationship between categories and techniques used in the laboratory and those used in the clinic. Reacting to Bourlière's presentation on the use of physiological measures to understand ageing processes across species, Olbrich, from the Geriatric Unit of Sunderland's General Hospital, asked,

> Do you know from which diseases your animals suffer or suffered, Prof. Bourlière, and for how long? Do you know the frequency of illness in your animals? In human beings we do know. Do you include all diseases or illnesses interfering with lifespan and, perhaps, also with processes of ageing? The process of ageing is a doubtful business altogether. Until now neither biologists nor scientists could present any evidence saying: 'These are the signs and symptoms of ageing only and not of a disease.' [...] The point is that you don't know the diseases. We know the human diseases, we know what humans die of, we have the post-mortems. Your wild animals have various diseases that you don't know; you have no post-mortems. So your vital statistics are not the vital statistics we deal with. Ageing must be something you can define and you, as a scientist, must be able to tell me as a clinician, 'These are the clinical signs of ageing.' And it must not be a disease; once it is a disease it is no longer 'ageing' but 'accident of death'.
>
> (Wolstenholme and O'Connor, 1957: 56–57)

At the time working within the Centre de Gérontologie de l'Association Claude Bernard – at the Paris Faculty of Medicine – Bourlière was by 1950 a highly respected physiologist, having written the first French language textbook of gerontology with his mentor Leon Binet – himself considered the true continuator of Bernard's approach to physiology as applied to medicine (Binet and Bourlière, 1955). Aiming to identify the 'common denominator' that would provide the basis for a biological definition of ageing, Bourlière's presentation drew on clinical syndromes – well described in physiology – such as arteriosclerosis, and proposed that future investigations of the 'metabolic rate' should be of use in clinical approaches to ageing. Olbrich, a practising physician, objected to this direction of research because it did not enable communication between clinicians and scientists. He argued that instead of dwelling in the 'doubtful business' of researching ageing by seeking 'common denominators', biologists should seek to provide him with a clear boundary between ageing and illness. This required that they worked with the same materials and instruments (post-mortems, vital statistics, clinical semiology, etc.) to identify equivalences between what was found in the clinic and the physiology laboratory.

This call for clinically applicable criteria would have seemed old-fashioned to Bourlière, trained as he was in Bernardian physiology. However, it also pointed towards an ambivalence of biologists of ageing with focusing on what they saw as 'the negative side of ageing' – i.e. its entanglement with illness and the institu-tional devaluation of older people in work, healthcare, etc. As we saw in Chapter 6, around this same time Nathan Shock and his colleagues at the Gerontological Branch had welcomed the opportunity to shift their research from disease-burdened populations towards 'healthy' older individuals to establish what came to be known as the Baltimore Longitudinal Study of Aging. Indeed, the key motivation behind most of the longitudinal studies of ageing launched in that decade had been to provide a fuller, more balanced, picture of the process of ageing than the one captured solely in geriatric clinics. Bourlière himself, in com-munity studies conducted from the late 1950s, had focused much of his work on comparing quantitative physiological measures of ageing in different populations, conceived as 'natural experiments' in gene-environment interaction (Bourlière and Parot, 1962). The promise of biological gerontology appeared to be that by identifying the normative, physiological parameters of the ageing processes, geri-atricians would be able to reinterpret the meaning of presenting symptoms.

This is usually taken to be broadly in concert with the aims of geriatricians themselves. From the 1940s onwards, particularly in the US but also in Europe (Korenchevski, 1961), geriatrics aimed to establish itself within the medical spe-cialities, chiefly by evoking an ideal notion of 'normal ageing' and then defining the expertise of the geriatrician as dealing with the prevention and treatment of diseases of old age (Hirshbein, 2000). But what Olbrich was highlighting was that this division of labour between the laboratory and the clinic missed a key infrastructural ingredient: a shared tool or measure of illness of old age, the 'symptoms and signs of ageing' as seen in the procedures and techniques nor-mally used by working clinicians. Without this, geriatricians and biologists were condemned to speak disparate languages, deal with different populations and live in different sociotechnical, biosocial worlds.

It was to address this problem that, from the 1960s onwards, gerontologists began floating the idea of bringing together all the strands of research on ageing under one roof. This proposal relied on the belief that 'different points of view' and approaches could be harmonised and coordinated under a shared programme of research. The template for this research policy approach had been established by the experience of the Medical Research Council in the UK from the 1920s to the 1940s, but particularly by the wider political vision underpinning the cre-ation of the NIH in 1946 in the US. All of those relied on the political recogni-tion of a public health problem – tuberculosis for the original MRC – to call into action an expert-led institutional framework for focused research. Although the 'problem of ageing' enjoyed similar recognition, it was uncertain whether a corresponding research organisation was required. It was not only that experts continued to disagree on what ageing was as an epistemic object, but also that the political framing of the ageing problem was differently enacted in different settings.

In the US, the process was facilitated by the fact that gerontological research had been integrated into the programme of the NIH since the latter's creation. Tellingly, such incorporation had not satisfied either mainstream clinical scientists or ageing researchers, leading to the establishment of the NIH NICHD in 1963 specifically to focus 'on the complex health problems and requirements of the whole person rather than on any one disease or part of the body' (see Chapter 6). A concern with the 'whole person' as opposed to disease-specific programmes was deployed as an explicit disaffection from the aims of the NIH, and its domination by academic physicians. However, this emplacing of gerontological research separated it from clinical institutions, and furthermore excluded most of the experimental research on cell biology of ageing (Park, 2016). This was not satisfactory either, and US gerontologists spent most of the 1960s lobbying for a federal programme that could 'coordinate research on the biological origins of aging' (Lockett, 1983: 85). Importantly, the drafting of legislation from 1968 onwards to create the desired federal programme or institute proved very controversial, including a presidential veto in 1972. This controversy was a compound of divergences between gerontologists and the medical establishment, and among gerontologists themselves.

When the National Institute of Aging was eventually formed in 1974, it represented a double compromise. On the one hand, it aimed to balance the interests of those invested in the 'whole person' and those concerned with a more restricted approach to ageing mechanisms. On the other, and most importantly, it intended to align the new institute with the contemporaneous social contract for science, and focus on 'deriv[ing] new knowledge to advance our understanding of the underlying causes of the aging process and help us separate disease from aging' (Butler, 1980: 4). However, when the new Director of the NIA suggested that 'research on aging has shifted from its exclusive disease orientation toward a more comprehensive investigation of the normal, physiological changes with age' (Butler, 1977: 8), the evocation of normality should be seen, given what was argued above, as more immediately related to the political need to rearticulate how Americans viewed older citizens' role in society, than to the needs of clinicians working with older people. Indeed, communication channels with clinicians, policy makers and the public were only fully established through the NIA's sponsoring of clinical research on a new bioclinical entity – Alzheimer's disease – from the early 1980s onwards, as we will explore in the next section.

In the United Kingdom, a different configuration of medicine, biology and old age not only distanced gerontology from biology altogether, but resulted ultimately in a significant weakening of gerontology. As in the US, support for ageing research was initially privately sourced, as a result of Lord Nuffield's personal interest in, and support for, Vladimir Korenchevski's research on the endocrinology of ageing from 1940. This programme received a significant boost with the publication of the National Insurance Act of 1946, leading to its sponsoring of the Survey Committee on the Problems of Ageing and the Care of People. The programme subsequently expanded throughout the 1940s and 1950s into an impressive network of laboratories across British universities, funding both basic research on ageing, such as Maynard Smith's experiments on longevity on the fruit fly (e.g. Smith, 1958), and gerontology applied to the industrial setting (see Chapter 7). Indicative of this confidence in the alignment between public interest and research was the fact that Peter Medawar, Britain's iconic scientist of the mid twentieth century, chose to focus his inaugural lecture on being appointed to a chair in the zoology department at University College London on the topic of ageing. His view was that,

> Everyone now knows that the proportion of older people in our population is progressively increasing. The economic consequences of such an age-structure are all too obvious. Now, biological research is by no means uninfluenced by the economic importunities of the times, and there can be little doubt that the newly awakened interest of biologists in ageing – or the hard cash that makes it possible for them to gratify it – is a direct reaction to this economic goad.
>
> (Medawar, 1952: 6)

Cast within a narrative where science was 'pulled' by the economic realities of the age to concentrate on ageing, Medawar's lecture can be seen as an attempt to regain some control of the direction of research in that domain. In particular, it proposed that an evolutionary understanding of ageing should be able to frame the more burning 'problems of ageing' (Moreira and Palladino, 2008). His theory, drawing on the work of Fisher (1930), was that, in an optimal population, the reproductive value of organisms should be inversely correlated with the probability of death, thus pushing an assortment of deleterious genes to regions of the lifespan that are only experienced by domesticated animals and 'civilised man'. However, Medawar's evolutionary model of ageing led paradoxically to the view that there was no reason to believe that there was a 'common denominator' driving the ageing process, senescence being 'simply [the] name for that whole group of causes which make animals have a determinate life-span instead of an indeterminate one' (Comfort, in Wolstenhome and O'Connor, 1957: 242). This meant that ageing research was not underpinned by a shared empirical referent, an ontological reference point, but instead by conventional, worldly valuations of ways of living.

The impact of Medawar's vision for ageing research should not be underestimated, as his trajectory through the scientific establishment before and after receiving the Nobel Prize for Medicine in 1960 marks him as one of the key figures in the shaping of biomedical research in the UK. His influence was pivotal, for example, in persuading the Medical Research Council to fund Alex Comfort's laboratory from 1964 onwards, when the Nuffield Foundation discontinued its programme. However, he was also influential in the MRC's decision to withdraw its funding to Comfort just a few years later (CMFT: Medawar Letter to MRC, April 1969). His view that experimental ageing research in the UK was of weak quality impacted significantly on MRC leaders, who used this judgement to avoid creating a coordinated programme of research on ageing in the 1970s or 1980s.

This path was further facilitated by the fact that, in the UK, the main link between ageing research and the clinic lay outside the laboratory. As Martin has observed, it was not through the laboratory but 'through the technique of the survey, [that] doctors created a body of knowledge relating to the social, economic, and medical needs of the aged population in their own districts' (Martin, 1995: 458). In the process, British geriatricians came to define gerontology as the speciality, in Lord Amulree's definition, concerned with 'those elderly sick with social and economic problems' (Amulree in Martin, 1995: 460). This in effect positioned the biology of ageing and gerontology as a research field outside the hospital, the main research platform of British biomedicine during the second half of the twentieth century (Stewart, 2008). Conversely, the association between old age and that peculiarly British disciplinary approach that went by the name of 'social medicine' was responsible for much uncertainty around the place of the elderly within the NHS.

Under these circumstances, any funding for research on the medical problems posed by the elderly tended to be allocated to disease-specific programmes within the MRC because, following Medawar's vision, there seemed to be nothing biologically and clinically distinctive about old age. Consequently, in the late 1970s, when social medicine lost its precarious institutional support within both the MRC and the NHS (Porter, 1997), and geriatrics faced increasing recruitment and status challenges, British gerontology all but lost its residual disciplinary legitimacy. Indeed, the separation between gerontology and clinical medicine was so wide that Comfort could propose, without challenge, that while 'the whole tendency of medicine, of public health – whether social or political – and of all the social progress which has been made in most countries [was] to produce a squarer and squarer curve', the assignment of ageing research was instead 'an *entirely different one*' (Comfort, 1968: 97–100, my emphasis). This meant that epistemic, normative and institutional factors worked together to progressively disconnect biological explanations of ageing from any public debates and programmes to address the 'problem of old age', such that by the end of the 1970s Peter Medawar himself could declare British gerontological research as moribund (Medawar and Medawar, 1977).

In France, the process of emplacement of gerontology in the research system is entangled with the history of the formation and development of the Association Claude Bernard, referred to above. Created in 1952 with the aim of delivering Bernard's vision of a close integration between research and clinical practice by funding full-time researchers to be integrated within Paris hospitals, by 1956 it included a Centre of Gerontology led by François Bourlière at the Sainte-Perrine Hospital. Focused on what Bourlière defined as the 'ecology of human ageing' (Bourlière, 1960: 314), it complemented the experimental physiological research developed with Binet, to include extensive measurement of community-dwelling participants' health, psychological abilities and living conditions in different socio-economic contexts. With the fusion of the Institut National d'Hygiène and the Association Claude Bernard into the Institute National de la Santé et de la Recherche Médicale (INSERM) in 1964, it was expected that the integrative impulse of the Association would cause the emerging recognition of French genetics and molecular biology (e.g. Monod) to benefit French medicine (Picard and Mouchet, 2009: 131–148).

Bourlière's gerontology unit was, however, not integrated into the new national institute, mainly because its research – on physiological ageing – was not seen to fit any of INSERM's 13 scientific sections. It was only in 1973, when the Institute launched a variety of key, priority 'thematic areas' that the Unité de Recherches Gérontologiques (INSERM U. 118) was created. Headed still by Bourlière, it included teams working on cell biology of ageing, endocrinology, ethology and 'ecology of ageing'. Its integration into the institute continued however to be a difficult business, with consistent discussion and uncertainty on where to institutionally and geographically locate it, and with consistent negative evaluations of its multidisciplinary outputs (INSERM: Bourlière, U118 1978 Report: 1–2). As in the US and Britain, the unwillingness of gerontologists to 'medicalise' ageing was partly responsible for this uncertainty. Although France, as a population experiencing one of the fastest rates of ageing, certainly had the political backing necessary to establish a coordinated programme of research, its development took place mainly outside the clinic, a process that was paradoxically reinforced by the creation of INSERM U. 118, with no clear links to clinical practice. This meant that even researchers focused on areas with obvious potential links to 'the problems of ageing', such as the lead investigator in a project on the endocrinology of menopausal change in the ovary, could publicly yearn for 'exchanges of views with gynaecologists' (INSERM: Aschheim, U118 1978 Report: 34).

In sum, if the 'problems of ageing' emerged during the years between the late 1930s and the 1970s as a pressing political question and a variety of powerful institutions became interested in the biology of ageing, the successful alignment of the 'problem of ageing' with biology and medicine was a highly contingent affair. While a variety of programmes and projects had been established which provided a secure footing for the development of ageing research, the articulation between these and clinical practices was less than straightforward. At the core of this mismatch was gerontologists' commitment to rescue ageing from the professional and institutional influence of medicine and from its entanglement with pathology that had been forged during the late nineteenth and early twentieth centuries (Katz, 1996: 40–48). As a result, ageing researchers and most particularly biologists of ageing struggled, until the 1970s, to communicate with clinicians and to enrol them in their sociotechnical networks. This was all about to change.

Hybridising ageing, making Alzheimer's disease

It is generally accepted that the reawakening of interest in senile dementia in the 1960s was sparked by the publication of studies led by Martin Roth and colleagues at Newcastle-upon-Tyne (UK), which correlated the number of neuritic plaques in patients' brains with the scores obtained by those patients in cognitive tests (Roth *et al.*, 1966). Developments in electron microscopy in the early 1960s had fostered a redescription of the neuropathological features of dementia at the ultra-structural level and this created interest in neurobiology among neuropathologists. This interest reshaped Alzheimer's disease during the 1970s, and was the basis for a number of etiological theories that were proposed in that decade, the most important of which addressed the toxic effects of aluminium in the brain and a deficit of the neurotransmitter acetylcholine. This last hypothesis, supported by neurochemistry studies that linked the cholinergic system in the brain and the cognitive deficits observed in patients suspected to have Alzheimer's disease, became the focus of a considerable proportion of the Alzheimer's disease research in the late 1970s and early 1980s (Perry *et al.*, 1978; Whitehouse *et al.*, 1982).

Importantly, these advances in the understanding of the biology of the disease were coordinated with an intensive process of characterisation of the disease processes from a clinical/behavioural perspective. Already in synchrony with the Newcastle correlation studies, Blessed, Tomlinson and Roth had developed an informant-based instrument to assess memory, concentration and orientation (Wilson, 2014). This was followed by a series of tools aiming to measure mental status, such as the Mini Mental Status Exam (Folstein *et al.*, 1975), tests concerned with 'clinically observable deterioration', and others aimed at assessing behavioural changes, or cognitive performance. This coordination was immensely facilitated by the establishment of the Alzheimer's Disease Research Centres by the NIA in 1984, and supported by the newly published diagnostic criteria (McKhann *et al.*, 1984).

The publication of what came to be known as the 'McKhann criteria' can, from this perspective, be seen as a major turning point in the establishment of Alzheimer's disease as a bioclinical entity. Developed through a consensus conference organised by the National Institute of Neurological and Communicative Disorders and Stroke and the Alzheimer's Disease and Related Disorders Association (later Alzheimer's Association), with support from the NIA, the workgroup advanced a vision of the integration between research, therapeutic experimentation and clinical practice. Recognising that there was 'insufficient knowledge about the disease', it admitted the criteria were 'tentative and subject to change' (McKhann *et al.*, 1984: 939). To address these uncertainties, the workgroup proposed that diagnostic of 'possible' or 'probable' Alzheimer's disease required a harmonisation between clinical, neuropsychological and laboratory investigations that should be used in the clinic and further research on the condition. In this regard, the McKhann criteria can be adequately conceived as a 'conventional standard' (Cambrosio *et al.*, 2009) in that those involved in setting the criteria engaged in explicit reflexive deliberations about the equipped and provisional nature of knowledge production. Importantly, they focused on describing how existing techniques and tools could be used to identify a new illness, thus enabling the transit of cases and materials such as brain tissue between laboratories and clinics. Their concern was not the description of 'normal ageing' from which pathological states could be derived; instead, the McKhann criteria provided a temporary demarcation of a bioclinical, hybrid area of interest.

There can be no doubt this strategy was successful in establishing Alzheimer's disease as one of the key domains of biomedical research from the late twentieth century onwards. In this process, no small part was played by the NIA, particularly after the nomination of Zaven Khachaturian as Director of the Neuroscience and Neuropsychology of Aging Program in 1977, which enveloped these constituencies, political actors and the 'American public' within what Robert Butler called the 'health politics of anguish' (Fox, 1989). The 'politics of anguish' and the activities of the Alzheimer's Association were also key in the NIA's efforts to obtain budget increases from the US Congress. With the support thus secured, the NIA experienced an influx of researchers from other areas of biomedical research, an influx also encouraged by 'aggressive recruitment', as one informant once has put to me. This process helped to transform the NIA role in the American polity but only in the form of a disease-specific programme that was to deploy the Institute's majority of resources (Ballenger, 2006). In this context, it was possible for this bioclinical collective to establish itself in the public arena, with further assistance from expert calculations of the dimension of the 'problem of dementia' in the US (e.g. Katzman, 1976), now being recognised as one of the main causes of death in developed countries.

This shift in approach to the research agenda on ageing was not without its critics, most publicly Richard Adelman, who denounced it as a process of 'Alzheimerization of ageing' (Adelman, 1995). That Adelman's critique, which had started a few years later when standing as President of the Gerontological Society of America, was used by Estes and Binney as evidence supporting their biomedicalisation of ageing thesis should not surprise us. However, in so doing, Estes and Binney failed to link his critique to wider uncertainties relating to the knowledge base and institutional deployment of Alzheimer's disease. Because the fact was that the alignment between the worlds of research, clinical practice, politics and patient advocacy that underpinned the emergence of the bioclinical collective for AD was built upon shifting foundations. For example, the same molecular techniques that had first energised AD research in the 1970s were already, during the late 1980s, suggesting possible alternatives to the 'cholinergic hypothesis'. Also, by then, it was becoming clear that expectations, fostered during the 1970s and early 1980s, for a 'rational', unproblematic translation of the cholinergic hypothesis into safe pharmacology were unrealistic. When the results of clinical trials of cholinesterase inhibitors (ChEIs) started surfacing in the 1990s, the expectations in the clinical research community had been already significantly lowered (Moreira, 2009). This was further compounded by the evolving relationship between different types of dementia.

While the definition of AD proposed during the 1970s relied on its differentiation from the vascular models of dementia that had been popular before, new work demonstrated that vascular pathologies, notably atherosclerosis, white matter lesions and mid-life arterial hypertension, were associated with AD and could enhance cognitive loss. Furthermore, research on the biology of dementia with Lewy bodies, Parkinson disease dementia and fronto-temporal dementia led to a redefinition of the classification of the dementias. In this classification, AD shared characteristics with both amyloidopathies, such as Familial Amyloid Polineuropathy (Corino de Andrade's Disease), and taupathies, such as fronto-temporal dementia, or progressive supranuclear palsy. These trends, it was increasingly realised, could potentially lead to the disaggregation of the 'identity' of the AD and its bioclinical collective.

Those concerned with linking research and clinical diagnosis and management of AD had to deal with different problems. While the establishment of the McKhann criteria helped to consolidate and stabilise the category of AD itself, the striving towards reliability and consistency across AD clinics may have produced a paradoxical effect. From their inception, clinical assessment tools such as the Global Deterioration Scale had included milder-than-dementia levels of cognitive impairment and seemed to suggest a continuous path in this condition. The introduction of standardised criteria for the diagnosis of AD in 1984, however, excluded persons presenting with 'mild memory problems'. Thus, in the next ten years there was a multiplication of terms to categorise the 'forgetfulness' experienced by a growing number of patients perhaps affected by increased public awareness of the cognitive symptoms of AD. This reopened a longstanding controversy about whether AD is qualitatively different from normal ageing or quantitatively different along a cognitive continuum, which is worth exploring.

Re-hybridising Alzheimer's disease

When conducting research on the epistemic and political transformation in the field of AD in the 2000s, I interviewed a variety of international experts to identify key processes and drivers in those changes (Moreira et al., 2008). While some saw the uncertainties described in the last section as exciting challenges deriving from newly formulated biomolecular models of dementia (e.g the amyloid cascade hypothesis), others saw the emergence of a more complex picture for the aetiology of dementia as a return to key questions that had been ignored so as to secure the AD definition (Moreira and Bond, 2008).

As one senior UK epidemiologist told me, changes in the AD field were mostly visible in,

> the exponential increase in the volume, the fact that it's become a fashion-able and attractive area to work in, and I think that's a lot to do with cre-ating the entity of Alzheimer's disease, making it something that was defined and it was okay to be labelled with it, and as I say, fashionable – it had money attached to it so people started wanting to work in it as an area, so that's been one of the positive outcomes for this delineation of it as something very different from ageing, but then more recently the big change, well, it's not really a big change, it's a reinvent – it's a reawaken-ing of the awareness of different factors and different continua, [...] an awareness that the field is more complex. I think people in the nineties had a feeling that they could crack it with the discovery of a single gene, and there was perhaps a period of a few years of a great hope in that place – within the community that saw it as a simple – a simple disease with a simple mechanism, clearly defined and therefore we ought to be able to crack it easily, and I think now people understand it's much more complex and heterogeneous.
>
> (Interview with SE, May 2005)

Her analysis was that while the delineation of AD as 'something very dif-ferent from ageing' had facilitated the influx of researchers and funding into the area, as discussed in the previous section, this had been done through an over-simplification of the causes of the onset of the illness. Describing first the reali-sation by the research community of the role of 'different factors and continua' as a big change, she then provides a historical narrative whereby this recognition is cast as a return to a lost awareness of complexity (Lock, 2013). Based on her own experience of the evolution of the field, this statement constitutes what could be labelled a members' 'critical history' of AD (Thévenot, 2009), identify-ing paths not taken, overlooked possibilities and biased decisions. The import-ance of such histories should not be underestimated, as they provide important clues to how actors are attempting to shape contemporaneous sociotechnical net-works (Moreira, 2000).

Reconstructing the largely implicit critical history presented in this and other interviews entails looking back to the formative years of Alzheimer's disease. As Robert Butler (1999) would later recall, the alignment between the NIA and the community of researchers interested in Alzheimer's disease was marked by the organisation of the 1977 Workshop on Alzheimer's Disease-Senile Dementia and Related Disorders by Robert Katzman, Robert Terry and Kathryn Bick. This workshop is seen as important because it brought together most researchers active in the field of senile dementia, particularly in the Anglo-Saxon world, and included Roth, by then at the University of Cambridge. Interested in a wide variety of topics that included anxiety and hallucination, his invitation was predicated on the studies he had conducted in the 1960s, which relied on linking clinical signs of senile dementia, scores obtained by participants in cognitive scales and neuropathological analysis (see above). The organisers of the workshop were convinced that the Newcastle studies presented compelling evidence of there being a pathological entity which could be described as Alzheimer's disease. Wanting to make this belief consensually visible, Leon Sokoloff, a professor of pathology, asked Roth,

> DR SOKOLOFF: That brings us around to the last question, is it a disease or is it normal aging?
>
> DR ROTH: Would you accept senile dementia as a disease?
>
> DR SOKOLOFF: Yes.
>
> DR ROTH: If you accept senile dementia as a disease, then the existing evidence, such as it is shows that it is merely an extreme variant of what you find in the brains of perfectly well-preserved people who go to their graves with plaques in their brains, nobody being the wiser. If it is an extreme variant of that sort, there may be hope of influencing the factors that potentiate that change, and there are many factors which do: pernicious anemia, hyperthyroidism, head injury, and perhaps serious cardiac failure. If you accept that this is so, then we are justified in investigating the conditions under which the accumulation keeps you this side of the threshold.
>
> (in Katzman *et al.*,1978: 265–266)

This exchange is revealing because it contrasts a purely qualitative differentiation between disease and normal ageing with a quantitative, nuanced, Bernardian approach that views dementia as an 'extreme variant' along a continuum. While Sokoloff embraced the strategy of coordinating research through a focus on a bioclinical entity, Roth was proposing an investigation that emphasised the control of the conditions that enable maintenance of cognitive abilities. It's not that Roth thought that there was no relation between plaques and senile dementia, but he believed this was part of a multifactorial association rather than a direct causal link. Thus, one key issue underpinning the divergence between the NIA-sponsored researchers and Roth was that while the former looked to the clinic to delineate the new pathological entity, Roth and colleagues had put their clinical observation of variations in 'old age mental disorder' to test in the community.

This divergence between the clinic and the community is of significance because it stands for the two different approaches to assembling ageing as an epistemic and political object, which have permeated the debates and controversies discussed in this chapter. One makes the clinic what classic Actor-Network theorists would call an 'obligatory passage point' (Callon, 1986) in the biomedicalisation of ageing. This relies on working with and transforming the expertise of clinicians in diagnosing and managing age-associated diseases. The other, as Wilson has shown in relation to the development of the Blessed scale (Wilson,

2014), is aimed at suspending the authority of the clinician to find new norms and standards. It uses the population as a laboratory (see Chapter 6), and looks to establish a 'true' value from averages and distribution of factors. It is not only that these approaches enact ageing in different ways, it is also that they embody conflicting views on how to organise institutions, practices and materials to manage ageing. This is hinted at in Roth's reply to Sokoloff, in his suggestion that prevention rather than cure was the best method to deal with dementia. As the interview with the British epidemiologist provided above suggests, Roth's view and approach was not taken into account and became a marginalised perspective in the international field of dementia research, practice and policy. Such was the sense of exclusion, that, during the 1980s, dementia epidemiologists felt it was impossible to publish and discuss data that did not confirm the discrete nature of Alzheimer's disease.

Reacting against this state of affairs, two of Roth's Cambridge collaborators, Brayne and Calloway, portrayed it as 'dogmatic' belief that they were set to dispute in a paper in the *Lancet* in 1988. In it, they argued that,

> In the community, 'normal ageing', benign senescent forgetfulness and Senile Dementia of the Alzheimer's Type do not fall into discrete categories but appear to lie on a continuum. Although the groupings are useful for planning treatment, the cut-off points are arbitrary.
>
> (Brayne and Calloway, 1988a: 1265–1267)

Brayne and Calloway had used scales to test the boundaries of the clinical category and instead of a bimodal distribution they had found a continuum. However, they recognised the need for 'groupings' for clinical practice. It was just that Brayne and Calloway qualified them as 'arbitrary' categories. This, in effect, insulated clinical judgement from a direct relation with the evidence-base and, most importantly, from the collective production of knowledge on ageing. This questioned not only the knowledge base of Alzheimer's disease but also its knowledge making arrangements, as described in the previous section.

The methodological assumptions of their research were quickly questioned by researchers invested in the Alzheimer's disease entity. For example, epidemiologists working on the Rotterdam Study of Ageing suggested that the continuum was an artefact of the scales used:

> The scales are crude and probably subject to measurement error, and this might blur a true bimodality. [...] Even if we accept the unimodal distribution, [this] does not necessarily imply that the underlying pathophysiological changes are normal.
>
> (Hofman *et al.*, 1988: 226–227)

The critique hinged on the lack of sensitivity of the measurement tools used but also on the difference between neuropsychological scales and pathological changes in dementia. Without correlating their findings with neuropathological examination of brain tissue, it was impossible to check, the Dutch researchers argued, whether the continuum hid 'underlying' qualitative differences. Recognising that they were accused of ignoring the lessons of the Newcastle studies, Brayne and Calloway responded that,

> The scales were used because they have been correlated with Alzheimer's type pathology in previous studies. Hofman *et al.* may be correct that SDAT is distinct from normal ageing, but we would argue that the onus of the proof is on them.
>
> (Brayne and Calloway, 1988b: 514–515)

The controversy had reached an impasse, as both conceded that neither position had been falsified by the data. The results were seen as heavily reliant on the instruments used and on the assumptions made by the instruments about dementia as an object, its ontology. Furthermore, such assumptions and their methodological workouts related closely to how either side approached the relationship between knowledge production on ageing and which actors, institutions and practices they saw as responsible for its management. It was a matter of normative as well as epistemic and ontological disagreement. This was made clear in the *Lancet* editorial which tried to draw lessons from the controversy:

> The practical importance of the distinction lies solely in our response to the classification. If society assumes that 'disease' is preventable, treatable and requires rehabilitation and care but that normal ageing does not, then the difference becomes critical. [...] By contrast, if a decision is made that manifestations of normal ageing should be given any effective and efficient management available, the entire argument becomes irrelevant.
>
> (Anon, 1989: 477)

The question of whether Alzheimer's was a qualitatively different pathological entity or a 'threshold' within a continuum, and, we might add, how research was organised accordingly, was, according to the *Lancet*, one to be decided by 'society'. Although the invocation of 'society' was largely rhetorical, it laid down clearly the two alternative paths for the ageing biomedicalisation assemblage. One relied on distinguishing disease from ageing and focusing on describing, preventing and treating 'diseases of old age', while the other proposed a more encompassing strategy of management of 'normal ageing'. What is remarkable about this editorial is that it exactly identified the two divergent strategies used by researchers and policy makers in the next decades.

Again, as the interview quoted at the beginning of this section indicates, biomolecular models, neuropathological studies and the epidemiology of late-onset AD suggested a slow and etiologically complex progression of disease. Addressing these challenges, AD researchers advocated what they sometimes labelled as a 'paradigm shift' from a focus on reversing the symptoms of AD to a focus on the prevention of the onset of illness (Moreira, 2010; Lock, 2013). As research groups and companies became increasingly interested in finding pharmacological agents that would target the molecular mechanisms that precede neuronal death (amyloid aggregation, etc.), one of the strategies to implement this 'preventative paradigm' was the creation of new risk categories that could bridge between normal ageing and AD. Categories such as Mild Cognitive Impairment (MCI) were thus proposed to identify a population for research on the bioclinical antecedents of dementia and to test the effectiveness of preventative therapies for AD.

Proposed as a 'biomedical platform' (Keating and Cambrosio, 2003) rearticulating between different types of laboratories – molecular biology, neuropathology, neuropsychology, neuroimaging, etc. – and the clinic, MCI worked mainly as an epistemic scaffold in exploring the validity of biomarkers of AD within clinical and therapeutic research (Moreira *et al.*, 2009). As such, it provided a linkage between biomolecular techniques, neuropathology staging systems and commonly used clinical dementia scales, sedimenting hybrid 'risk categories' at the boundary between normal and pathological ageing. This approach is evident, for example, in a clinical trial in progress since 2012, which aims to use the effects of beta-amyloid immunisation in dominantly inherited AD as a model for the sporadic, late-onset type. Using heredity to identify participants, investigators proposed to trace amyloid deposition as measured in Positron Emission Tomography scans (PET) and cerebrospinal fluid analysis arguing that the 'biomarker approach is particularly important in this study as most study subjects will be cognitively normal at the time of enrolment and most will remain cognitively normal during the first 2 years of the study' (www.clinicaltrials.gov/ct2/show/record/NCT01760005).

Although crucial for investigators to draw equivalences and conserve passageways to more easily available – and accepted – standards such as the Clinical Dementia Rating scale (see above), this strategy is challenging because PET and/or biomolecular analysis of beta-amyloid are only available in research or selected clinics. This means that only if those circuits and networks of circulation are in action is it possible to maintain the AD bioclinical collective. There is evidence, however, that those circuits are fragile and depend significantly on the form of healthcare organisation in which clinicians are working (Moreira *et al.*, 2008). On the other hand, there are signs also that pharmaceutical companies have in the last years become less interested in the promises of neuroscience (www.pharmafile.com/news/172099/brain-drain), and are finding it hard to coordinate the large, longer clinical trials that are required by a disease-modifying, preventative approach to AD (Moreira, 2009).

In making those uncertainties and difficulties visible and public, a significant part has been played by those opposing the disease-specific approach to ageing research. Indeed, the arguments put forward by those wanting to focus on ageing as an underlying process rather than solely on its pathology can be said to constitute a critique of the epistemic and institutional foundations of biomedicine. In the next section, I analyse the emergence and consolidation of this critique.

Biogerontology as a critique of biomedicine

So far in this chapter I have explored how AD came to be constituted a hybrid bioclinical entity supporting new collaborative networks between laboratories, clinics, patient associations and policy makers around the problem of ageing, where previous attempts to do so had failed. However, I have also described how this bioclinical entity became increasingly fragilised by developments in the epidemiology of ageing, new molecular models of the disease and a difficult 'translation' of the preventative paradigm into new therapeutic approaches. The significance of these uncertainties is in large part a function of the degree of stability that was previously ascribed to AD as a form of 'pathological ageing', whereby concerned actors realised – to use Marx and Engels' expression – that 'all that was solid melted into air'. Such a transformation is also indicative of the role of AD as a paradigmatic case for wider debates about the normative aims and sociotechnical organisation of contemporary medicine and how it relates to the 'problem of ageing', as evidenced by Estes and Binney's use of the illness to support their biomedicalisation of ageing thesis.

For a significant proportion of biologists of ageing, the consolidation of AD represented a path that neglected the complexity of senescence processes at the level of the cell and organism (Adelman, 1995). As the uncertainties relating to the aetiology of AD surfaced at the turn of the century, biologists of ageing saw this as an opportunity to mount a critical attack on the application of the 'biomedical model' to ageing. Their argument, however, was not that this process represented solely a distortion of the ontological character of ageing, but more importantly that it entailed a series of normative contradictions and epistemic tensions. It was, in this regard, the hybrid nature of biomedicine that was at stake. In this process, biogerontologists attempted to turn their lack of connection with clinicians, which had hampered the stabilisation of their concepts and tools since the 1930s, as the basis for proposing a new approach to ageing and health.

The pivotal issue in this critique concerns what critics see as the dependence on clinical definitions of those diseases most commonly associated with old age. Revisiting the debates between geriatricians and biologists in the 1950s (see above), gerontological critics of biomedicine argue that the focus on the symptomatic pathological states risks ignoring what Bourlière had once called the 'common denominator' linking ageing, health and illness. The question is, again, related to the distinction between normal and pathological, with which we started this chapter. The clinical worldview, these biogerontologists maintain more specifically, was well-suited to pathologies characterised by discrete and specific aetiologies, but is inadequate to address the chronic, long-term illnesses of the late twentieth and early twenty-first centuries. Their charge is that the temporal unfolding of these illnesses is so nearly correlated with ageing that it unsettles the epistemic pairing of the 'normal' and the 'pathological' that underpins the clinic. Focusing on the pathological aspects of the ageing process obscures understanding the diverse and complex processes involved in senescence and ageing more generally. As one internationally respected biogerontologist and practising geriatrician once put to me in an interview:

> Some people call [age-associated diseases] ageing, some people call that normal ageing, [and] some people say it's different from normal ageing. I don't make a distinction between them. [...]
>
> What's behind, let's say the senile muscle, is of equal importance [as Alzheimer's disease] because people can't go out any more and they suffer from it. What's normal? There's nothing like normal ageing.
>
> [We] were not meant to live longer than forty years, and the system is optimised in an environment like Africa. But now we don't live in Africa....
> [Our] life history is ... the result of ... an old, optimised genome ... now ... exposed under modern, affluent conditions. But it's not meant to be ... it's not meant to be exposed under these conditions.

Herein, lays the crucial boundary between clinicians and biogerontologists, as defined by the latter: they don't make a distinction between normal and pathological. To undermine this normal–pathological dichotomy, they, like my interviewee, draw on an evolutionary, Darwinian tradition of reasoning – first applied to ageing by Medawar (see above) – that locates the structuring of the human genome within the environment that gave rise to the species. In this perspective, the human organism is optimally adapted to conditions most of us no longer experience, having access to a variety of technological innovations (agriculture, urbanisation, etc.) that protect the organism and enhance longevity. Ageing is thus the 'result of an old, optimised genome exposed to affluent conditions'. Normality, as defined by clinicians, is thus seen as a historically specific discrimination of little biological significance, thus making the category of 'normal ageing' an arbitrary, historically contingent label. However, it is not as if biogerontologists think that categories do not have consequences.

Indeed, the critical work of biogerontologists has been focused on denouncing both the causes and consequences relying on categories such as normal ageing. For example, one reading of a position statement signed by some of the most important international contemporary biogerontologists (Butler *et al.*, 2008) is that they object to the reliance on the clinic to understand and manage ageing because it represents a historical contradiction, where nineteenth century models of disease aetiology are imposed on twenty-first century institutions. This condemnation of the biomedical model is also succinctly encapsulated in an interview with Robin Holliday by a fellow biogerontologist and his former research student, Suresh Rattan,

> SR: But do we need to do research on each disease separately or do we need to develop a common research programme under the framework of the biological basis of ageing?

RH: That is the way most doctors look at disease. Whenever there is any particular disease, they tend to treat that specific disease, and ignore the fact that other parts of the body may be failing. Of course the biomedical field is so huge that each specialist is an expert in only one disease, and if something else is wrong a different specialist is called in. In my opinion there needs to be a whole re-appraisal of the field of age-related disease, and much more emphasis on ageing research. What is important is to look for the origin of each disease and then try to prevent it from happening.

(Rattan, 2002: 320)

For Holliday and his fellow biogerontologists, the problem with conventional medicine was its compartmentalisation of diseases, and the assumption of there being a specific cause for each identified illness (the doctrine of specific aetiology). This correspondence between cause and disease, which had its origins in hospital medicine in the nineteenth century, and relied on the analytical work of collecting and classifying pathological specimens, had been challenged by the emergence of experimental medicine, often linked as we saw above to the figure of Claude Bernard, and more recently to the emergence of the preventative medicine paradigm. As a result, they argue, contemporary biomedicine cannot but fail to deliver the treatments of conditions commonly associated with old age, because it continues to insist on treating diseases separately. This entails a politics of expertise that segments knowledge and practice, a form of disaggregation that goes against what was known about the 'origin of disease', as the case of AD demonstrated.

Claiming to embody the force of reason and scientific progress, critics of biomedicine, like their social science counterparts, looked for professional power, commercial interests, the political establishment, or a combination of these, to explain the present situation. Miller (2002), for example, noted that 'senators' and voters' parents [die] of specific diseases', reinforcing the organisation of biomedicine and the focus of research funding. Similarly, the US Alliance for Aging Research (2005), have blamed that combination of normative and epistemic commitments as responsible for the minute proportion within the annual budgetary allocation for the NIH attributed to ageing. Through this politics, as the interview quoted above makes clear, problems that 'are of equal importance' become hierarchised, focusing attention on 'illnesses' such as Alzheimer's disease to the detriment of frailty as an effect of the organisation of medical specialities and the illness markets they sustain.

In this critique, biogerontologists also make clear what alternative world they would like to bring to bear. In this world, where the evolutionary perspective is recognised (Nesse and Williams, 1996), health, its maintenance and management, is the main focus of research, practice and policy. This entails understanding how the very mechanisms that ensure our survival are turned into threats, under specific conditions (Holliday, 2007): there lies the 'origin' of disease.

Such a configuration of the problem requires a wholesome transformation of the system of biomedical research and innovation. Instead of disease-specific programmes, which biogerontologists see as responsible for slowing the rate of effective biomedical innovation since the 1970s, there should an emphasis on basic ageing research (Miller, 2009). As the House of Lords Science and Technology Committee noted, 'generic research into the process of ageing ... may be the most direct route to developing novel interventions and therapies' (2006). This is because, according to the Alliance for Ageing Research, 'the aging research field [itself is] on the threshold of a new way of thinking – shifting from a focus on specific age related illnesses to a search for an understanding of aging itself' (Alliance for Ageing Research, 2005: 4).

Understanding 'ageing itself' is in effect a rhetorical token to signify the stabilisation of an institutional alignment of evolutionary models and genetic research on ageing from the 1980s onwards. Although there is considerable uncertainty and debate on the effect of evolutionary processes on genetics, this alignment is powerfully embodied in Thomas Kirkwood's work at the turn of the 1980s. Kirkwood rearticulated Medawar's and 'first generation' evolutionary explanations of ageing, based on a fixed, Mendelian concept of the genome, by combining molecular and demographic analyses to advance the notion that the organism should be understood as the product of a process involving the balancing of energetic investments – trade-offs – between maintenance of the germ line and the somatic body (Kirkwood, 1977). Those trade-offs were an optimal evolutionary solution aimed at maximising the chances of successful reproduction of the germinal line, which would have to be balanced against the energy cost of these investments to the continuity of this same line. On this evolutionary understanding of ageing, attention is directed towards the molecular mechanisms involved in the preservation of genomic integrity, or, as Kirkwood has put it, towards 'the evolved capacity of somatic cells to carry out effective maintenance and repair' (Kirkwood and Austad, 2000: 235).

It is this linkage between evolution, molecular dynamics and health that defines the business of gerontology as enhancing the ability of the individual to 'carry out effective maintenance and repair'. In this, and importantly, gerontology ceases to be a field of research specialisation concerned with the 'problems' of a distinct population – the elderly – as these 'problems' are rearticulated as unfolding temporally on to antecedent risk factors and biomolecular pathways. Biomolecular and demographic pathways of ageing individuals who might be 'at risk' of developing diseases such as Alzheimer's or cardiovascular disease are traced backwards to the earliest possible genetic, molecular, behavioural or clinical manifestations with the aim of developing multiple preventative interventions. These diseases then become part of a wider set of 'degenerative diseases' that are only connected contingently to the organism's chronological age. Within this configuration of gerontology, all degenerative diseases might be said to entail 'ageing'.

An example of this strategy is the application for the FDA to authorise researchers to conduct a clinical trial of Metformin – a drug licensed for the treatment of Diabetes II – to test its ability to prevent a plethora of other diseases. As *Nature* reported at the time,

> Doctors and scientists want drug regulators and research funding agencies to consider medicines that delay ageing-related disease as legitimate drugs. Such treatments have a physiological basis, researchers say, and could extend a person's healthy years by slowing down the processes that underlie common diseases of ageing – making them worthy of government approval. [...] Current treatments for diseases related to ageing 'just exchange one disease for another', says physician Nir Barzilai of the Albert Einstein College of Medicine in New York. That is because people treated for one age-related disease often go on to die from another relatively soon thereafter. 'What we want to show is that if we delay ageing, that's the best way to delay disease.'
>
> (Hayden, 2015: 265)

Because the domain of gerontology thus defined is resistive to any precise delimitation, it can become an object of interest for all clinical practitioners involved in managing degenerative diseases, from the primary care practitioners controlling their patients' hypertension to the specialised clinicians required to train these practitioners in the assessment of the earliest symptoms of illness. In this respect, biogerontology, as the biomedicalisation of ageing theorists have suggested, offers opportunities of development to a great variety of actors in the market for health care. An investigation of the mechanisms involved in the onset of illness greatly expands opportunities for companies because the threshold of treatment moves ever backwards to encompass a greater fraction of the population. In so doing, biogerontology centrally aligns the technoscientific promises of the 'ageing society' with the institutions and procedures of the bioeconomy, linking the added value of biotechnological innovation to health care insurers' expectation of reducing the prevalence of degenerative diseases so as to reduce the aggregate cost of provision. This is a crucial point.

This redefinition of gerontology draws on evolutionary biology to allow explorations of organisms' life histories in relation to genes and environment, so enabling links between the laboratory, preventative medicine and health maintenance programmes. In an important positions statement already referred to above, key biogerontologists, including Kirkwood, have argued for a renewed articulation of these key institutions of contemporary biomedicine suggesting that,

> the exploration of the mechanisms by which ageing can be postponed in laboratory models will yield new models of preventive medicine and health maintenance for people throughout life, and the same research will also inform a deeper understanding of how established interventions, such as exercise and healthy nutrition, contribute to lifelong wellbeing.
>
> (Butler *et al.*, 2008: 399)

As Palladino and I (2008) have argued, while this proposal aligns biogerontology with wider transformations in the field of medicine and health care identified by Clarke and her colleagues (see above), as discussed in the first section of the chapter, it does so by explicitly challenging the role of the clinic within biomedicine and its normative, ultimate goals. This is related to how they imagine health screening and management programmes to work, bypassing for the most part the close pastoral guidance of biomedical experts to mediate between the laboratory and everyday life. This is because the clinics' and the clinicians' focus on managing illness is, from their perspective, infrastructurally inadequate to deliver their vision of health and wellbeing. Instead, they propose that standards, such as biomarkers of ageing (see Chapter 5), will enable the monitoring of individuals' ageing processes and the evaluation of the role of interventions in 'lifelong wellbeing'.

In this regard, contemporary biogerontologists can be said to be reinventing the proposals articulated by Alex Comfort at the turn of the 1970s (also Ludwig, 1980). Like Comfort, they see the clinic as an obstacle rather than a mediator in the implementation of the biology of ageing. Like him, they actively differentiate between contemporaneous medicine, the aim of which is to 'produce a squarer and squarer survival curve', where illnesses are postponed to the later stages of life (see Fries, 1980), and their project, which aims to 'move the whole [curve] to the right by an unspecified amount' (Comfort, 1968: 100). This alternative project views the relationship between mortality and morbidity as essentially modifiable, and resists arbitrary limitations on the worth and length of human life which Comfort labelled as 'misreadings of biology of human societies [resulting from] a compound of illiberal opinion and bad science' (Comfort, 1964: 271).

This genealogical trace has, in my view, been responsible for the clustering of biogerontologists with more radicalised versions of the modifiable nature of the human life span, such as the prolongevity movement (Gruman, 2003; Anton, 2013), or anti-ageing medicine itself, which biogerontologists have vigorously resisted (Binstock et al., 2006). Because, while most contemporary biogerontologists agree that there is a natural limit to human longevity, written in the 'old genome' my interviewee referred to, they also share an understanding of the intimate relationship between health and longevity, which impacts on average life span in the whole population. Again, from their perspective, what the status quo defines as the human life span is the result of contingent historical conditions, and, they argue, while there might be a natural limit to longevity, there is uncertainty on what exactly that is for humans. Most importantly, biogerontologists, like Comfort, define themselves as anti-status-quo but do not proselytise their aims and objectives beyond the usual network of policy makers. They are, in this regard, not a social movement in the same way as one could see anti-ageing as being (cf. Mykytyn, 2006).

This reluctance to extend their policy network combined with the strategy of double differentiation, in relation to clinicians, on the one hand, and to the anti-ageing movement on the other, while defining the whole epistemic identity and authority of biogerontology is also its Achilles heel. By relying on networks of policy makers, usually assembled in expert discussions and seminars such as the ones organised regularly by the International Longevity Centre (www.ilc-alliance.org/), biogerontologists are dependent on contextual policy cycles. One key problem in this is that it is difficult to identify biogerontology's affinitive social constituency, its public. By expanding the relevance of the process of ageing to the whole of the life course, and linking it to health maintenance and wellbeing, biogerontology speaks to most if not all citizens of contemporary, technological societies, as it directly addresses their 'biological citizenship' (Rose and Novas, 2005). That is to say, it speaks to all but not to any one group in particular. Diverging from twentieth century gerontology's close relationship with older people's advocacy aims and organisations (Moody, 2006), however, biogerontology has not been able to construct an elective affinity with any group's claims to justice or recognition.

Indeed, biogerontologists' fragile capacity to transform biomedicine as they intend might be one of the reasons they are seen by some social scientists as belonging to the biomedicalisation assemblage they intend to criticise. From the perspective of social scientists, biogerontologists' refusal of normal/pathological distinctions and their separation from clinical worlds of practice will appear as a minor detail, a variation on the same process of expansion of the health management systems across domains of social life (Fishman *et al.*, 2010). My view is that their critique of the scientific, technological, institutional and normative basis of biomedicine presents a unique challenge and an alternative to medicine as we know it and adds to our understanding of the complexly diverse ways in which ageing, health and illness are articulated in contemporary societies. They present a critique of the 'biomedicalisation of ageing' while bringing it firmly into the technoscientific horizon of the 'ageing society'.

Conclusion

In this chapter, I have argued that to understand the way in which the ageing society is deployed through technoscientific promises it was necessary to focus on the attempt to apply biological models and techniques to the ageing process. I have suggested that to do so requires bringing together two disparate understandings of the process of biomedicalisation: one that emphasises the reciprocally constitutive process by which biomedicine and ageing become entangled, and another that exactly problematises the basis on which this entanglement is seen to take place. This refers to a dynamic whereby attempts to undo clinical categories of normal and pathological ageing have been challenged by the need to support communication and exchange between the laboratory and the clinic through the creation of hybrid bioclinical 'platforms'.

I have demonstrated that this dynamic is still at work in the making and unmaking of the bioclinical entity of Alzheimer's disease, in that while it enabled material and symbolic exchanges between clinicians, researchers and patients, this was reliant on the closing off of uncertainties about the boundaries and multiplicities of dementia. Opening these uncertainties, contemporary bio-gerontologists have articulated a technoscientific vision that challenges not only Alzheimer's disease but also the very institutional and ontological basis of bio-medicine. In so doing they have positioned themselves between mainstream medicine and the anti-ageing movement, and their institutional alliances have as a result been fragile at best, except when deployed within the new configuration of the bioeconomy. As we have seen in the first section of the chapter, however, this reliance on commodification and consumerism rubs uncomfortably with the need that biogerontologists have to establish clinical partnerships, innovations such as telomere length measuring being simultaneously defined as 'disruptive innovation' and 'just like' other forms of health risk screening. These ambiguities and indecisions are indicative of the uncertainties still affecting the material, epistemic and normative basis on which the technoscientific promises of the ageing society, particularly its focus on health, can be brought to bear.

Chapter 10

The end of the 'ageing society'?

This book's point of departure was that there is a need to recognise how science and technology have contributed to defining and articulating ageing as a societal issue. While the official version of the 'ageing society' is that it results solely from the interaction between demographic changes and modern social and economic institutions, for which new technological and scientific solutions are needed, in this book I have argued that science and technology have significantly shaped the way in which we understand and manage the relationship between 'population problems', the economy, society and individuals. My proposal was to set up a research programme that delves deep into the knowledge making practices, the tools, the technologies, the devices, the knowledge making socio-technical institutions that underpin the ageing society.

My second proposal was that if we aim to focus on the epistemic infrastructure of the 'ageing society', what we find is uncertainty and multiplicity – most notably evident in the controversy between 'pessimistic' and 'optimistic' analyses of the societal consequences of demographic ageing. To explore these uncertainties and multiplicities, the book investigated a diverse – multiple, coexisting – array of 'agencements' that are key to understanding the relationship between science, technology and the 'ageing society'. But this approach had another objective: to avoid an uncritical search for ways to restore and secure agency for older people within technological democracies. This is particularly important, I argued, in a context where there are increasing calls for ageing individuals to actively engage with science, technology and medicine in order to enhance health, activity and their overall involvement in social, economic and cultural spheres. Rather than juxtaposing present, unsatisfactory circumstances with idealised forms of getting older, I suggested that it is possible not only to show how 'it could have been otherwise' but also – through the deployment of patchwork stories – to bring to bear the ways in which it has always been otherwise, that is to say, to fully exhibit the 'heterogeneity of what was once imagined consistent with itself'.

I provided the first piece of the patchwork in Chapter 3 where I detailed the epistemic character of the 'ageing society' by linking its assembling – its gathering into a thing – to an emerging collective inquiry into the actuarial techniques and practices upon which the liberal welfare state had been based. First outlining a genealogy of the relationship between the 'population problem' and the economy, the chapter suggested that the 'social security crises' of the early 1980s brought to bear concerns about the accuracy and reliability of demographic calculations and economic forecasting to accurately predict increased life expectancies and correlated imbalances of social security programmes. The 'ageing society' can only, from this perspective, be described as an established uncertainty concerning how to deploy procedures of scientific research and technological innovation in addressing and managing these uncertainties.

I suggested further that the public recognition of the epistemic – but unstable – underpinnings of the 'ageing society' served as the context to rearticulate a set of technoscientific promises from the 1980s to the present day. This relied on a reconfiguration of ageing as a matter of concern (Latour, 2005) – i.e. on a rearticulation of how to intertwine knowledge production and political process in relation to ageing. In as much as the 'insurance society' had relied on and reinforced models of ageing that enacted 'old age' as a fixed biological, psychological and socioeconomic stage in life (Katz, 1996), for which specific clinical and social security programmes were required, this rearticulation required *both and simultaneously* opening up established knowledge on age-related functional decline – on the relationship between disability and death – and exploring what kinds of institutions would be most suitable to new arrangements of 'vitality'. The uncertainties attached to this double exploration were, I suggested, partially settled through the pronouncement of a set of technoscientific promises offering to modify health and work through an assortment of converging tools and forms of knowledge – a series of agencements – ranging from the molecular to the sociological.

In Chapter 4, I explored the role of what I labelled the eugenic agencement in the stabilisation of the link between the 'mechanisms of life' and the economy. I argued that this epistemic and moral concern with 'the quality of populations' and migratory flows played a significant part in shaping the social security systems of liberal welfare states, including old age pensions. Tracing how population science experienced a shift away from lowering birth rates towards a focus on life and health expectancy, I suggested that this transformation was achieved through a reinforcement of the problematic role of migration in the cultural economy of population management. In this, the possible quantitative redress of population ageing by migration became increasingly trumped by the aim to bolster the use of existing 'human capital', through the extension of health and active working life.

From this perspective, the technoscientific promise of raising the efficiency and productivity of existing human capital was the result of a series of epistemic and policy lock-in – and lock-out – processes. Partially in line with what Rose (2009) proposed, this resulted in a shift in the logics of biopower, whereby the classification of the eugenic value of persons is replaced by a valuation of the genetic and molecular components of ways of living. However, my view is that this relied on opening up death to biopower, rather than its replacement by a sole focus on morbidity. Drawing on models of epidemiological transition, I showed how management of the factors associated with the onset of mortality in the life course defined the 'central issue of vitality' in populations in the 'ageing society'. This also means that the relationship between illness and death remains an open, unstable matter within the 'ageing society'.

Chapter 5 explored how age measurement was identified as the infrastructural key to the double exploration of the links between ageing, illness and death, on the one hand, and the institutions that would be most suitable to new arrangements of 'vitality', on the other. I have proposed that, in this process, a variety of epistemic and normative uncertainties about the role of chronological age to ascertain age-specific norms, values and expectations were publicly raised. In order to resolve these problems, a multiplicity of 'personalised' age standards were proposed between the 1940s and the 1980s. These were grouped in four types. First, there were those aimed at using functional assessments to enhance efficiency. The second type included measurements that linked the monitoring of somatic qualities to health and efficiency. Third were measures that linked functionality to the unique contribution of older individuals to society. Lastly, there were measurements focused on exceptional longevity.

Because none of these alternative standards has been able to replace chronological age, I have argued that these two processes have combined to produce torquing effects between chronological age and individual trajectories, and between and within alternative age measurements such as 'biological age'. Such torque effects were explored in more detail in relation to biological age, which can be seen as the dominant alternative to chronological age in contemporary societies. In relation to this, I described how uncertainties in biological age validation and implementation have drawn on and reinforced two modes of organising knowledge production to meet the challenges of the 'ageing society': one, that relies on chronological age as a variable to create clusters of health risk – and on biomedical experts as pastoral guides of individual conduct (Rabinow and Rose, 2006) – and another which aims to embody in the alternative standard itself the machinery of individualisation of health, work and technology.

Chapter 6 further investigated the issue of individualisation in the ageing society. Taking as a point of departure the established consensus that ageing is a process where genetic, environmental, lifestyle and experiential factors combine in unique individual trajectories, it found – like the previous chapter – fundamental uncertainties about the meaning and use of the individual as an epistemic entity to understand and manage ageing. I proposed that the best window into these uncertainties was an in-depth exploration of a methodological agencement – the longitudinal study of ageing – which I conducted through a case study of the genesis and development of the BLSA from the 1950s until the present. This exploration showed how the BLSA's evolution was underpinned by an interaction between different sets of epistemic practices, devices and norms, particularly as understanding the relationship between ageing, health and life expectancy became more pressing in wider society.

This dispute is partially connected to the divergence encountered in relation to biological age. On the one side, there were those who stressed the importance of randomisation and sampling in grounding disease management procedures. These were based on age group averages and regressions to the mean. They were what could be labelled as Queteletian in their approach to ageing and illness, aiming to use population data to calculate a 'true value' – usually described as 'normative' – of the ageing process. Here, individuals are seen to result from the distinguishing combination of shared characteristics, along multiple variance curves. On the other side, there were those who evoked the epistemic repertoire of physiological research, especially the notion of the 'model organism' – a purified, simplified workable exemplar, normally contained within laboratory-like conditions. I located these practices in the articulation between 'biological age' and 'uniqueness' described in the previous chapter, in that unusual, extraordinary cases were viewed as epistemically valuable – as epistemic indexes – exactly because they lacked 'representativeness'.

Again, a similar reliance on exemplary models was identified in relation to enactments of the concept of functional age in Chapter 7. The chapter's focus was on the epistemic and political formatting of later life through active and healthy ageing instruments, a type of agencement that is specifically focused on the instrumental value of work. Taking the WAI as an example of these agencements, I traced its linkages to functional age measurements concerned with understanding work as a function of the internal dynamics of body-in-action. This was particularly visible in how the focus on military work – the soldier, the aviator, etc. – between the two world wars shaped the research of industrial psychologists, gerontologists and ergonomists interested in the factors that underpin the biological and psychological self-regulating capacities of labour.

Crucially, I identified how the concept of functional age was itself a compound of two different enactments of the problem of ageing and work. One favoured the procedures and norms of physiological research, and again used model organisms – such as the Olympic athlete – to identify the fundamental relationships shaping work capacity and fatigue. It used the exercise physiology laboratory as the index setting to enact these relationships and to articulate a vision of the 'optimal fit between man and machine'. The other was the Cambridge approach, aimed at identifying the 'key elements of skill' by abstraction from representative samples of the population. They used realistic simulations of ordinary work situations to generate that data, but were dependent on their previous expertise on light, skilled military labour to make claims on ageing at work. This bounded expertise was a key factor in decreasing Cambridge's influence, as the economy transitioned to a post-industrial organisation, where the ageing society's emphasis on technoscientific promises of health reignited interest in stress and 'optimal adaptations'.

Such focus on 'optimal adaptation' was also present in the reconfiguration of the role of home and place in the care of older people. Thus, in Chapter 8, I argued that it was through articulating the promises of using technology in the home and community that the complex dynamics between ageing-in-place – usually enacted as safety and security of older people – and the autonomy and independence denoted in active ageing practices was made visible. I first explored how a social and political critique of the caring relationship embedded within the welfare state led to a re-evaluation of the role of extended families and communities in the care of older people. This enabled an acknowledgement of the care recipient as an autonomous, rather than necessarily dependent, person. However, this line of collective investigation also made autonomy a mutable quality, reliant upon the environment in which older people live, an artificiality it shared with the type of subjective rationality that Foucault identified as distinctive to neoliberal dispositifs of power.

I suggested that this linkage was enacted by tools such the IADL, an example of a type of agencement of the psycho-physical capacities of 'prosthetic environments' for older people. Tracing the assemblage around the tool, I argued that its analytical decomposition of daily life – cooking, housekeeping, laundry, etc. – transposes an experimental form of reasoning to the home, whereby familiar objects are recast as items of 'control' or 'conditioning'. There is however continuing and unresolved debate about whether technology should be used in a 'therapeutic' approach or merely as a compensation for loss of function, as these construct different political scaffolds for ageing-in-place. I argued further that this agencement's formal decomposition of older people's relationship with their 'environments' is constantly challenged by the way in which care is crafted in the way persons accommodate themselves to their surroundings through trial and error, and questions the sustainability of the regime of technoscientific promises upon which the ageing society is founded.

This questioning continued in Chapter 9. In this chapter, I proposed that an analysis of the application of biotechnology to ageing requires an acknowledgement that the current scientific and commercial investment in age-retarding interventions is intimately associated with the reorganisation of health care around technological intervention and the modes of prevention and consumption. However, it is also important to realise that these processes, while significant, remain at the margins of mainstream biomedicine. I suggested that this was because biologists of ageing have been unable to establish and stabilise circuits of material and symbolic exchange with clinicians, including geriatricians. Where this was possible, as in the case of Alzheimer's disease, it relied on a fragile boundary between normal and pathological ageing, a boundary that became undone as understanding of the illness progressed.

Part of this undoing was the work of biogerontologists themselves, who saw it as an opportunity to vindicate their alternative model of understanding and managing diseases that are commonly associated with age, such as heart disease, stroke and cancer. Reconstructing these diseases as part of a wider set of disorders linked to cellular senescence, biogerontology continues however to struggle to establish the hybrid, collaborative platform of knowledge production that is characteristic of innovation in the life sciences. It has thus relied mainly on circuits outside mainstream clinical medicine to expand its networks. As a result, techniques such as telomere length measuring are being simultaneously defined as 'disruptive innovation' and 'just like' other forms of health risk screening, becoming ambiguously linked to technological modification of the ageing process that is however commercialised only as a 'novel food'. These ambiguities and indecisions are indicative of the predicaments affecting the technoscientific promises of the ageing society.

By tracing the multiplicities and the partial connections between different agencements within the 'ageing society' my key aim was, as noted above, to open up the 'heterogeneity of what was once imagined consistent with itself'. In so doing, I have also brought into view the fragilities and contradictions of the technoscientific promises attached to the 'ageing society'. They are reliant on particular enactments of health and its relationship with mortality, of work and activity, and of technology's capacity to modify or reassemble ageing processes. These I described as a series of conditionally obligatory lock-ins and lock-outs: of emphasising science and technology as solutions rather than constituent parts of the 'problem'; of framing the enhancement of 'vitality' by reinforcing the 'border machine'; of imagining new forms of age measurement by undermining modern democratic institutions; of referring to an idealised unique individual but enacting it mainly through population matrices; of embedding a narrow functional enactment of health in collective imaginaries of wellbeing; of promising independence and autonomy through the deployment of conditioning domesticating techniques; of transforming health into a technoscientific project without fully articulating what this means for social and economic institutions in the future.

One view on the consequences of the contradictions and heterogeneities of particular forms of sociotechnical organisation – capitalism, democracy, the 'ageing society', etc. – is that they contain the seeds of their own demise. This is particularly evident in the Marxian tradition of the social sciences. From this perspective, this book would constitute a critical denunciation of the weaknesses of the 'ageing society'. Identifying and analysing these inconsistencies would enable assertions and claims about the 'end of the ageing society', and perhaps what could be expected afterwards. A slightly different perspective, which this book espoused, is that such contradictions are inherent to the transformation of sociotechnical collectives. Exploring heterogeneities, in this regard, is not ultimately aimed at smoothing differences and seeking equivalences between divergent worlds in a new formation, but instead at investigating 'the unknown constituted by these multiple, divergent worlds, and the articulations of which they could eventually be capable' (Stengers, 2005: 995; Law, 2015; Moreira, 2012a). Tracing multiplicities is thus an attempt to renew the sense of the 'unknown' by opening up 'what seemed consistent with itself'.

Rather than announcing the demise or unmanageable conditions of the 'ageing society' the book aimed to shift the premises and assumptions on which the debates on ageing are mostly conducted. In particular it argued, from a social science point of view, that our commitment to recuperate agency for older people might be best served by focusing on empirically exploring and proliferating the range of possible worlds, of agencements – of distributions of agency – that belong to the assembling of the ageing society. This should enable the multiplication of points of entry, of forms of participation in the collective. Some of these might not look like agency and empowerment, but by tracing the partial connections between different distributions we are also opening possibilities, and investigating the unknown.

This means that what this book offers is a starting point for this exploration of possibilities. It urges other social scientists or concerned groups to further multiply the sociotechnical traces of the 'ageing society'. This will be a task focused on remaking our engagement with key agencements of the 'ageing society', and in so doing, aiming to transform the way in which knowledge making practices approach the ageing question. As I suggested, the 'ageing society' is first and foremost an impending uncertainty, which was partially stabilised by the articulation of technoscientific promises. The uncertain character of the 'ageing society' is an opportunity for us to further unsettle its foundations. In this, it is my view that focusing on the three axes of technoscientific promise of the ageing society would be of the utmost importance.

First, there is a need to undertake a detailed investigation of the multiple enactments of health that have emerged since the 1940s. It will be particularly important to understand how the recognition of the inaccuracy of actuarial predictions – discussed in Chapter 3 – impacted on the search for new measurements of health, and how those embodied different versions of how health should be managed in society (Moreira, 2012a). This will entail going beyond the accepted typology of health concepts that divided them into functional, naturalistic or holistic, and developing a grasp of how health measurement formats relate to specific normative and political worlds and, how indeed, the same form of health measurement can contain divergent worlds within. For example, an ongoing investigation into the evolution of self-rated health from the 1950s to the present suggests that it is implemented both as an independent variable to explain mortality, a mediating factor to explain health service use and a dependent outcome of the quality of health services. This diversity is indicative of a multiplicity of forms of valuing health – and of enacting the ethic of the 'good life' – that require analysis.

Another dimension that should be in this research agenda concerns the use of functional age assessments in retirement procedures in organisations and in decisions to withdraw from the labour force. As the definitions of function and health overlap, researchers will be interested in tracing and understanding how models of 'optimal adaptation' rub uncomfortably with politicised versions of disability rights within work organisations. Conversely, it would be interesting to explore how emerging versions of health as 'optimal adaptation' (Huber *et al.*, 2011) have reignited the debate about the role of medicine and doctors in setting the norms for social life. This might entail reinventing occupational medicine or reconfiguring the collaboration between work organisations and clinical or public health services.

Third, there is the need to continue the important work being developed in the ethnography of the use of technology at home. While, for funding reasons, this has been mostly focused on 'telecare', new developments in robotics and ambient technology require a constant interaction with social science observation and analysis of the distributions of power and agency enacted in such uses. My suggestion is that these take seriously the 'politics of prosthesis' in technology for older people and how this might be in tension with the enactment of agency that underpins the imaginaries of the 'silver market': who purchases assistive technologies specifically designed for 'older people', and how are those purchases organised?

A similar line of questioning is suggested for further investigations into the 'biomedicalisation of ageing'. As more compounds – rapamycin etc. – move into the evaluation phase, it will be important to document and trace the ways in which experimental protocols might also contain social and organisational innovation on how to deliver and assess the effect of these technologies in real populations. On the other hand, as the rate of innovation in the field increases, it is necessary to deploy social science expertise in implementing models of responsible innovation, by helping concerned groups and publics interaction with scientists. For this, social scientists need to become embedded in a field which they have mostly critiqued from a distance. This should be crucial in helping rearticulate a more diverse, encompassing technoscientific regime for the 'ageing society'.

References

Abrams, P. and Bulmer, M. (1986) *Neighbours: The Work of Philip Abrams.* Cambridge: CUP Archive.

Aceros, J. C., Pols, J. and Domènech, M. (2015) 'Where is grandma? Home telecare, good aging and the domestication of later life', *Technological Forecasting and Social Change,* 93(1): 102–111.

Achenbaum, W. A. (1978) *Old Age in the New Land: The American Experience since 1790.* Baltimore: Johns Hopkins University Press.

Achenbaum, W. A. (1995) *Crossing Frontiers: Gerontology Emerges as a Science.* Cambridge: Cambridge University Press.

Adams, V. (2016) *Metrics: What Counts in Global Health.* Durham, NC: Duke University Press.

Adelman, R. C. (1987) 'Biomarkers of aging', *Experimental Gerontology,* 22(4): 227–229.

Adelman, R. C. (1995) 'The Alzheimerization of aging', *The Gerontologist,* 35(4): 526–532.

Adey, P. (2010) *Aerial Life: Spaces, Mobilities, Affects.* Oxford: Wiley-Blackwell, pp. 114–144.

AFAR (American Federation for Aging Research) (2011) *Biomarkers of Aging.* New York: AFAR.

Agamben, G. (1998) *Homo Sacer: Sovereign Power and Bare Life.* Stanford: Stanford University Press.

Agamben, G. (2009) *'What is an Apparatus?' and Other Essays.* Stanford: Stanford University Press.

Akrich, M. (1992) 'The de-scription of technical objects', in W. Bijker and J. Law (eds) *Shaping Technology/ Building Society.* Cambridge, MA: MIT Press.

Alliance for Aging Research (2005) The Science of Aging Gracefully: Scientists and the Public Talk about Aging Research. Available at: www.policyarchive.org/bitstream/handle/10207/5577/science_of_aging_gracefully.pdf (accessed 24 March 2009).

Amoore, L. (2006) 'Biometric borders: Governing mobilities in the war on terror', *Political Geography,* 25(3): 336–351.

Amsterdamska, O. (2005) 'Demarcating epidemiology', *Science, Technology & Human Values,* 30(1): 17–51.

Anderson, B. (1983) *Imagined Communities.* London and New York: Verso.

Ankeny, Rachel (2008) 'Wormy logic: Model organisms as case-based reasoning', in Angela N. H. Creager, Elizabeth Lunbeck and M. Norton Wise (eds) *Science without*

Laws: Model Systems, Cases, Exemplary Narratives, pp. 46–58. Chapel Hill: Duke University Press.

Ankeny, R. A. and Leonelli, S. (2011) 'What's so special about model organisms?' *Studies in History and Philosophy of Science Part A*, 42(2): 313–323.

Anon (1989) 'Editorial', *Lancet*, 333(8636): 477.

Anton, T. (2013) *The Longevity Seekers: Science, Business, and the Fountain of Youth*. Chicago, ILL: University of Chicago Press.

Arber, S. and Ginn, J. (1991) *Gender and Later Life: A Sociological Analysis of Resources and Constraints*. London: Sage.

Armstrong, D. (1995) 'The rise of surveillance medicine', *Sociology of Health & Illness*, 17(3): 393–404.

Armstrong, D. (2000) 'The temporal body', in R. Cooter and J. Pickstone (eds) *Medicine in the Twentieth Century*, pp. 247–259. Amsterdam: Harwood.

Arthur, W. B. (1989) 'Competing technologies, increasing returns, and lock-in by historical events', *The Economic Journal*, 99(394): 116–131.

Avramov, D. and Maskova, M. (2003) *Active Ageing in Europe* (Vol. 1). Luxembourg: Council of Europe.

Avramov, D. and Maskova, M. (2004) *Active Ageing in Europe* (Vol. 2). Luxembourg: Council of Europe.

Bacon, F. W., Benjamin, B. and Elphinstone, M. D. W. (1954) 'The growth of pension rights and their impact on the national economy', *Transactions of the Faculty of Actuaries*, 22(January): 265–415.

Baker, III, G. T. and Sprott, R. (1988) 'Biomarkers of aging', *Experimental Gerontology*, 23(4): 223–239.

Ballenger, J. (2006) *Self, Senility, and Alzheimer's Disease in Modern America*. Baltimore: Johns Hopkins University Press.

Barcroft, J. (1914) *The Respiratory Function of the Blood*. Cambridge: Cambridge University Press.

Bartlett, F. C. (1927) *Psychology and the Soldier*. Cambridge: Cambridge University Press.

Bartlett, F. C. (1932) *Remembering: An Experimental and Social Study*. Cambridge: Cambridge University Press.

Bartlett, F. (1951) 'The bearing of experimental psychology upon human skilled performance', *British Journal of Industrial Medicine*, 8(4): 209–221.

BBC (British Broadcasting Corporation) (2013) 'UK woefully underprepared for ageing society, say peers', 14 March 2013. Available at: www.bbc.co.uk/news/uk-politics-21773743 (accessed 10 November 2014).

Beaud, J. P. and Prévost, J. G. (1994) 'Models for recording age in 1692–1851 Canada: The political-cognitive functions of census statistics', *Scientia Canadensis: Canadian Journal of the History of Science, Technology and Medicine (Scientia Canadensis: Revue canadienne d'histoire des sciences, des techniques et de la médecine)*, 18(2): 136–151.

Beck, U. (1986) *Risk Society: Towards a New Modernity*. London: Sage.

Beck, U. (2001) 'Living your own life in a runaway world', *Archis*, 2(2001): 17–30.

Bell, C. and Newby, H. (1972) *Communities Studies*. New York: Praeger.

Benjamin, H. (1947) 'Biologic versus chronologic age', *Journal of Gerontology*, 2(3): 217–227.

Berg, M. and Mol, A. (1998) *Differences in Medicine. Unravelling Practices, Techniques, and Bodies*. Durham, NC; London: Duke University Press.

Berlivet, L. (2005) '"Association and causation": The debate on the scientific status of risk factor epidemiology, 1947–c.1965', in Virginia Berridge (ed.) *Making Health Policy: Networks in Research and Policy after 1945*, pp. 43–74. Amsterdam: Rodopi.

Beveridge, W. (1942) *Social Insurance and Allied Services*. London: HMSO.

Beveridge, W. H. (1943) *Pillars of Security*. London: George Allen & Unwin.

Biggs, S. (1997) 'Choosing not to be old? Masks, bodies and identity management in later life', *Ageing and Society*, 17(5): 553–570.

Binet, L. and Bourlière, F. (1955) *Précis de gérontologie*. Paris: Masson.

Binstock, R. H., Fishman, J. R. and Johnson, T. E. (2006) 'Anti- aging medicine and science: Social implications', in R. H. Binstock and L. K. George (eds) *Handbook of Aging and the Social Sciences*. Amsterdam/Boston: Academic Press/Elsevier.

Birren, J. E. (ed.) (1959) *Handbook of Aging and the Individual*. Oxford, England: University of Chicago Press.

Birren, J. E., Butler, R. N., Greenhouse, S. W., Sokoloff, L. and Yarrow, M. R. (1963) *Human Aging: A Biological and Behavioral Study*. Bethesda, MD: National Institute of Mental Health.

Blackburn, E. H., Greider, C. W. and Szostak, J. W. (2006) 'Telomeres and telomerase: The path from maize, Tetrahymena and yeast to human cancer and aging', *Nature Medicine*, 12(10): 1133–1138.

Boltanski, L. (1990) 'Sociologie critique et sociologie de la critique', *Politix*, 3(10): 124–134.

Boltanski, L. (2011) *On Critique: A Sociology of Emancipation*. Cambridge: Polity.

Boltanski, L. and Thévenot, L. (1999) 'The sociology of critical capacity', *European Journal of Social Theory*, 2(3): 359–377.

Boltanski, L. and Thévenot, L. (2006) *On Justification: Economies of Worth*. Princeton, NJ: Princeton University Press.

Bond, J. (1992) 'The medicalization of dementia', *Journal of Aging Studies*, 6(4): 397–403.

Bookstein, F. and Achenbaum, A. W. (1993) 'Aging as explanation: How scientific measurement can advance critical gerontology', in Thomas R. Cole, W. A. Achenbaum, P. L. Jakobi and R. Kastenbaum (eds) *Voices and Visions of Aging: Toward a Critical Gerontology*, pp. 20–43. New York: Springer.

Borrell, B. (2012) 'Lawsuit challenges anti-ageing claims', *Nature*, 488(7409).

Bouma, Herman, Fozard, James L., Bouwhuis, Don G. and Taipale, V. T. (2007) 'Gerontechnology in perspective', *Gerontechnology*, 6(4): 190–216.

Bourlière, F. (1960) 'Gerontological activities in France, 1957–1960', *Journal of Gerontology*, 15(3): 314–316.

Bourlière, F. (1970) *The Assessment of Biological Age in Man*. Geneva: World Health Organization.

Bourlière, F. and Parot, S. (1962) 'Le viellissement de deux populations blanches vivant dans des conditions très différentes', *Rev. Fr. d'Études Clin. Biol.*, 7(6): 629–635.

Bourlière, F., Clément, F. and Parot, S. (1966) '"Normes" de vieillissement morphologique et physiologique d'une population de niveau socio-économique élevé de la région parisienne', *Cahiers du Centre de recherches anthropologiques*, 10(1): 11–39.

Bowker, G. and Star, S. L. (1999) *Sorting Things Out: Classification and Its Consequences*. Cambridge, MA: MIT Press.

Brayne, C. and Calloway, P. (1988a) 'Normal ageing, impaired cognitive function, and senile dementia of the Alzheimer's type: A continuum?' *The Lancet*, 331(8597): 1265–1267.

Brayne, C. and Calloway, P. (1988b) 'Is Alzheimer's disease distinct from normal ageing?' *The Lancet*, 332(8609): 514–515.

Broberg, G. and Roll-Hansen, N. (1996) *Eugenics and the Welfare State*. Ann Arbor: University of Michigan Press.

Bruce, K. and Nyland, C. (2011) 'Elton Mayo and the deification of human relations', *Organization Studies*, 32(3): 383–405.

Brückner, H. and Mayer, K. U. (2005) 'De-standardization of the life course: What it might mean? And if it means anything, whether it actually took place?' *Advances in Life Course Research*, 9(2005): 27–53.

Bryson, D. (1998) 'Lawrence Frank, knowledge and the production of the social', *Poetics Today*, 19(3): 401–421.

Bulmer, M. (1996) 'Edward Shils as a sociologist', *Minerva*, 34(1): 7–21.

Busch, L. (2011) *Standards: Recipes for Reality*. Cambridge, MA: MIT Press.

Butler, R. N. (1963a) 'The facade of chronological age: An interpretative summary', *American Journal of Psychiatry*, 119(8): 721–728.

Butler, R. N. (1963b) 'The life review: An interpretation of reminiscence in the aged', *Psychiatry*, 26(Feb.): 65–76.

Butler, R. N. (1969) 'Age-ism: Another form of bigotry', *The Gerontologist*, 9(4): 243–246.

Butler, R. N. (1977) 'Research programs of the National Institute of Aging', *Public Health Reports*, 92(1): 3–8.

Butler, R. N. (1980) 'Introduction' in S. G. Haynes and M. Feinleib (eds) *Epidemiology of Aging*, Proceedings of the 2nd Conference (28–29 March 1977), pp. 1–5. Bethesda, MD: National Institute of Aging.

Butler, R. N. (1982a) 'The triumph of age: Science, gerontology, and ageism', *Bulletin of the New York Academy of Medicine*, 58(4): 347–60.

Butler, R. N. (1982b) 'Preface' in M. E. Reffand and E. L. (eds) Schneider *Biological Markers of Aging*, Proceedings of the Conference on Non-lethal Biological Markers of Physiological Aging (19–20 June 1981), pp. 1–3. Washington, DC: US Department of Health and Human Services, National Institutes of Health.

Butler, R. N. (1983) 'An overview of research on aging and the status of gerontology today', *Milbank Memorial Fund Quarterly*, 61(3): 351–361.

Butler, R. N. (1999) 'Is the National Institute on Aging mission out of balance?' *The Gerontologist*, 39(4): 389–391.

Butler, R. N. (2003) 'Sam Greenhouse's scholarly contributions to human aging studies', *Statistics in Medicine*, 22(21): 3281–3284.

Butler, R. N., Sprott, R., Warner, H., Bland, J., Feuers, R., Forster, M., Fillit, H., Harman, S. M., Hewitt, M., Hyman, M., Johnson, K., Kligman, E., McClearn, G., Nelson, J., Richardson, A., Sonntag, W., Weindruch, R. and Wolf, N. (2004) 'Aging: The reality biomarkers of aging: From primitive organisms to humans', *The Journals of Gerontology Series A: Biological Sciences and Medical Sciences*, 59(6): B560–B567.

Butler, R. N., Miller, R. A., Perry, D., Carnes, B. A., Williams, T. F., Cassel, C., Brody, J., Bernard, M. A., Partridge, L., Kirkwood, T., Martin, G. M. and Olshansky, S. J. (2008) 'New model of health promotion and disease prevention for the 21st century', *BMJ: British Medical Journal*, 337(7662): 149.

Bytheway, B. (2011) *Unmasking Age: The Significance of Age for Social Research*. Bristol: Policy Press.

Caestecker, F. (2000) *Alien Policy in Belgium, 1840–1940: The Creation of Guest Workers, Refugees and Illegal Aliens*. London: Berghahn Books.

Çalışkan, K. and Callon, M. (2010) 'Economization, part 2: Research programme for the study of markets', *Economy and Society*, 39(1): 1–32.

Callon, M. (1986) 'Some elements of a sociology of translation: Domestication of the scallops and the fishermen of St Brieuc Bay', in J. Law (ed.) *Power, Action and Belief: A New Sociology of Knowledge*. London: Routledge & Kegan Paul.

Callon, M. (1991) 'Techno-economic networks and irreversibility', in J. Law (ed.) *A Sociology of Monsters: Essays on Power, Technology and Domination*, pp. 132–161. London: Routledge.

Callon, M. (1992) 'The dynamics of technoeconomic networks', in R. Coombs, P. Saviotti and V. Walsh (eds) *Mapping the Dynamics of Science and Technology*, pp. 70–83. Oxford: Blackwell.

Callon, M. (1998) 'An essay on framing and overflowing: Economic externalities revisited by sociology', *The Sociological Review*, 46(S1): 244–269.

Callon, M. (1999) 'The role of lay people in the production and dissemination of scientific knowledge', *Science Technology & Society*, 4(1): 81–94.

Callon, M. and Latour, B. (1981) 'Unscrewing the big Leviathan: How actors macrostructure reality and how sociologists help them to do so', in K. Knorr and A. Cicourel (eds) *Advances in Social Theory and Methodology: Toward an Integration of Micro- and Macro-Sociologies*, pp. 277–303. London: Routledge & Kegan Paul.

Callon, M. and Law, J. (1995) 'Agency and the hybrid "collectif"', *The South Atlantic Quarterly*, 94(2): 481–507.

Callon, M. and Rabeharisoa, V. (2004) 'Gino's lesson on humanity: Genetics, mutual entanglements and the sociologist's role', *Economy & Society*, 33(1): 1–27.

Callon, M. and Rabeharisoa, V. (2008) 'The growing engagement of emergent concerned groups in political and economic life: Lessons from the French association of neuromuscular disease patients', *Science, Technology & Human Values*, 33(2): 230–261.

Callon, M., Lascoumes, P. and Barthe, Y. (2009) *Acting in a Uncertain World: An Essay on Technical Democracy*. Cambridge, MA: MIT Press.

Cambrosio, A., Keating, P., Schlich, T. and Weisz, G. (2009) 'Biomedical conventions and regulatory objectivity: A few introductory remarks', *Social Studies of Science*, 39(5): 651–664.

Campisi, J. and di Fagagna, F. D. A. (2007) 'Cellular senescence: When bad things happen to good cells', *Nature Reviews: Molecular Cell Biology*, 8(9): 729–740.

Canguilhem, G. (1978) *On the Normal and the Pathological*. Dordrecht: Reidel.

Carr-Saunders, A. M. (1922) *The Population Problem: A Study in Human Evolution*. Oxford: Clarendon Press.

Chudacoff, H. P. (1989) *How Old Are You? Age Consciousness in American Culture*. Princeton, NJ: Princeton University Press.

Clark, M., Czaja, S. J. and Weber, R. (1990) 'Older adults and daily living task profiles', *Human Factors: The Journal of the Human Factors and Ergonomics Society*, 32(5): 537–549.

Clarke, A. E., Mamo, L., Fishman, J. R., Shim, J. K. and Fosket, J. R. (2003) 'Biomedicalization: Technoscientific transformations of health, illness, and U.S. biomedicine', *American Sociological Review*, 68(2): 161–194.

Cochoy, F. (1998) 'Another discipline for the market economy: Marketing as a performative knowledge and know-how for capitalism', *The Sociological Review*, 46(S1): 194–221.

Coleman, D. A. (2000) 'Who's afraid of low support ratios? A UK response to the UN

Population Division report on "Replacement Migration"', in United Nations 'Expert Group' meeting, New York. Available at http://citeseerx.ist.psu.edu/viewdoc/download? doi=10.1.1.414.1391&rep=rep1&type=pdf (accessed 10 October 2014).

Coleman, D. A. (2002) 'Replacement migration, or why everyone is going to have to live in Korea: A fable for our times from the United Nations', *Philosophical Transactions of the Royal Society of London B: Biological Sciences*, 357(1420): 583–598.

Comfort, A. (1956) *The Biology of Senescence*. Oxford: Rheinhart.

Comfort, A. (1964) *Ageing, the Biology of Senescence*. London: Routledge & Kegan Paul.

Comfort, A. (1968) *The Conquest of Ageing*. Saskatoon: University of Saskatchewan.

Comfort, A. (1972) 'Measuring the human ageing rate', *Mechanisms of Ageing and Development*, 1(1): 101–110.

Comfort, A. (1977) *A Good Age*. New York: Crown Publishers.

Committee on an Aging Society (1985) *Health in an Older Society*. Washington, DC: National Academy Press.

Committee on an Aging Society (1988) *America's Aging: The Social and Built Environment in the Older Society*. Washington, DC: National Academy Press.

Committee on Finance US Senate (1983) *Trends in US Life Expectancy*. Washington, DC: US Government Printing Office.

Committee on Population Problems of the National Resources Committee (1938) *Problems of a Changing Population*. Washington, DC: US Government Printing Office.

Connelly, M. J. (2009) *Fatal Misconception: The Struggle to Control World Population*. Cambridge, MA: Harvard University Press.

Connor, S. (2012) 'Want to know how old you really are? Then join the queue'. The Independent, 27 December 2012.

Conrad, P. (2007) *The Medicalization of Society: On the Transformation of Human Conditions into Medical Disorders*. Baltimore: Johns Hopkins University Press.

Cooper, M. (2011) *Life as Surplus: Biotechnology and Capitalism in the Neoliberal Era*. Seattle: University of Washington Press.

Costa Jr, P. T. and McCrae, R. R. (1980) 'Functional age: A conceptual and empirical critique', in S. G. Haynes and M. Feinleib (eds) *Epidemiology of Aging*, Proceedings of the 2nd Conference (28–29 March 1977), pp. 23–46. Bethesda, MD: National Institute of Aging.

Costa Jr, P. T. and McCrae, R. R. (1988) 'Measures and markers of biological aging: "a great clamouring … of fleeting significance": An answer to W. Dean and RF Morgan', *Archives of Gerontology and Geriatrics*, 7(3): 211–214.

Cot, A. L. (2011) 'A 1930s North American creative community: The Harvard "Pareto Circle"', *History of Political Economy*, 43(1): 131–159.

Cowdry, E. V. (ed.) (1939) *Problems of Aging*. Baltimore: Williams & Wilkins.

Cowgill, D. O. (1974) 'The aging of populations and societies', *The Annals of the American Academy of Political and Social Science*, 415(1): 1–18.

Cumming, E. and Henry, W. E. (1961) *Growing Old*. New York: Basic Books.

Czaja, S. J. (ed.) (1990) *Human Factors Research Needs for an Aging Population*. Washington, DC: National Academies Press.

Czaja, S. J. and Barr, R. A. (1989) 'Technology and the everyday life of older adults', *The Annals of the American Academy of Political and Social Science*, 503(1): 127–137.

Dannefer, D. and Settersten, R. A. (2010) 'The study of the life course: Implications for social gerontology', in D. Dannefer and C. Philipson (eds) *The SAGE Handbook of Social Gerontology*, pp. 3–20. London: Sage.

Daric, J. (1946) 'Vieillissement démographique et prolongation de la vie active', *Population* (French edition), 1(1): 69–78.

Daric, J. (1952) 'Vieillissement de la population, besoins et niveau de vie des personnes âgées', *Population* (French edition), 7(1): 27–48.

Daric, J. and Girard, A. (1948) *Vieillissement de la population et prolongation de la vie active: Les résultats d'une enquête par sondage*. Paris: Presses Universitaires de France.

Davis, K. (1945) 'The world demographic transition', *The Annals of the American Academy of Political and Social Science*, 237(1): 1–11.

Davies, D. F. and Shock, N. W. (1950) 'Age changes in glomerular filtration rate, effective renal plasma flow, and tubular excretory capacity in adult males', *Journal of Clinical Investigation*, 29(5): 496–507.

Dean, M. (2015) The Malthus effect: Population and the liberal government of life, *Economy and Society*, 44(1): 18–39.

Deetz, S. (2003) 'Disciplinary power, conflict suppression and HRM', in M. Alvesson and H. Willmott (eds) *Studying Management Critically*. London: Sage.

Desrosières, A. (1991) 'How to make things which hold together: Social science, statistics and the state', in P. Wagner, B. Wittrock and R. Whitley (eds) *Discourses on Society*, pp. 195–218. Dordrecht Reidel: Springer Netherlands.

Desrosières, A. (2008) *Pour une sociologie historique de la quantification*. Paris: Presses de l'École des Mines.

DG ECFIN (Directorate General of Economic Affairs) (2012) *The Impact of Ageing on Public Expenditure: Projections for the EU25 Member States on Pensions, Health Care, Long Term Care, Education and Unemployment Transfers*. Brussels: European Commission.

Dickinson, B. (2011) 'Telomere Nobelist: Selling a "biological age" test', *New Scientist*, 2810(April): 17–18.

Dilnot, A. W., Kay, J. A. and Morris, C. N. (1984). *The Reform of Social Security*. Oxford: Oxford University Press.

Dirken, J. (1972) *Functional Age of Industrial Workers*. Groningen: Wolters-Noordhoff.

Economic Policy Committee and the European Commission (2006) *The Impact of Ageing on Public Expenditure: Projections For the EU25 Member States on Pensions, Health Care, Longterm Care, Education and Unemployment Transfers (2004–2050)*. Luxembourg: European Commission.

Economic Policy Committee and the European Commission (2015) *The Impact of Ageing on Public Expenditure: Projections For the EU25 Member States on Pensions, Health Care, Longterm Care, Education and Unemployment Transfers (2004–2050)*. Luxembourg: European Commission.

Ehrlich, P. R. (1968) *The Population Bomb*. New York: Ballantine.

Epstein, S. (2008) *Inclusion: The Politics of Difference in Medical Research*. Chicago: University of Chicago Press.

Estes, C. L. (1979) *The Aging Enterprise*. San Francisco: Jossey-Bass.

Estes, C. L. and Binney, E. (1989) 'The biomedicalization of aging: Dangers and Dilemmas', *Gerontologist*, 29(5): 587–596.

European Commission (1999) *Towards a Europe for All Ages – Promoting Prosperity and Intergenerational Solidarity*. Brussels: European Commission. Available at http://ec.europa.eu/employment_social/soc-prot/ageing/com99-22/com221_en.pdf.

Eurostat (2012) *Active Ageing and Solidarity Between Generations: A Statistical Portrait of the European Union 2012*. Luxembourg: Office of the European Union.

Ewald, F. (1991) 'Insurance and risk', in M. Foucault, G. Burchell, C. Gordon and P. Miller *The Foucault Effect: Studies in Governmentality*. Chicago: University of Chicago Press.

Fanu, J. L. (1999) *The Rise and Fall of Modern Medicine*. London: Little, Brown.

Featherstone, M. and Hepworth, M. (1991) 'The mask of ageing and the postmodern life course', in M. Featherstone, M. Hepworth and B. Turner (eds) *The Body: Social Process and Cultural Theory*, pp. 371–389. London: Sage.

Fernández-Ballesteros, R., Robine, J. M., Walker, A. and Kalache, A. (2013) 'Active aging: A global goal', *Current Gerontology and Geriatrics Research*, Vol. 2013: Article ID 298012.

Finch, J. and Groves, D. (eds) (1983) *A Labour of Love: Women, Work and Caring*. London: Routledge & Kegan Paul.

Fisher, R. A. (1930) *The Genetical Theory of Natural Selection: A Complete Variorum Edition*. Oxford: Oxford University Press.

Fishman, J. R. (2004) 'Manufacturing desire: The commodification of female sexual dysfunction', *Social Studies of Science*, 34(2): 187–218.

Fishman, J. R., Settersten Jr, R. A. and Flatt, M. A. (2010) 'In the vanguard of biomedicine? The curious and contradictory case of anti-ageing medicine', *Sociology of Health & Illness*, 32(2): 197–210.

Folk, G. E. (2010) 'The Harvard Fatigue Laboratory: Contributions to World War II', *Advances in Physiology Education*, 34(3): 119–127.

Folstein, M. F., Folstein, S. E. and McHugh, P. R. (1975) ' "Mini-mental state": A practical method for grading the cognitive state of patients for the clinician', *Journal of Psychiatric Research*, 12(3): 189–198.

Foucault, M. (1973) *The Birth of the Clinic*, trans. A. Sheridan. London: Tavistock.

Foucault, M. (1980) *Power/Knowledge: Selected Interviews and Other Writings, 1972–1977*, ed. Colin Gordon. London: Pearson Education.

Foucault, M. (1991) *The Foucault Effect: Studies in Governmentality*, (eds) G. Burchell, C. Gordon and P. Miller. Chicago: University of Chicago Press.

Foucault, M. (2003) *Society Must Be Defended: Lectures at the Collège de France*. New York: Picador.

Foucault, M. (2004) *Society Must Be Defended: Lectures at the Collège de France, 1975–76*, trans. David Macey. New York: Picador.

Foucault, M. (2009) *Security, Territory, Population: Lectures at the Collège de France 1977–1978* (Vol. 4). Basingstoke: Palgrave Macmillan.

Foucault, M. (2010) *The Birth of Biopolitics: lectures at the Collège de France, 1978–1979*. Basingstoke: Palgrave Macmillan.

Fox, P. (1989) 'From senility to Alzheimer's disease: The rise of the Alzheimer's disease movement', *Milbank Memorial Fund Quarterly*, 67(1): 58–102.

Fozard, J. L., Graafmans, J. A., Rietsema, J., Bouma, H. and van Berlo, A. (1993) 'Aging and ergonomics: The challenges of individual differences and environmental change', in K. Brookhuis, C. Weikert, J. Moraal and D. De Waard (eds) *Aging and Human Factors: Proceedings of the Europe Chapter of the Human Factors and Ergonomics Society Annual Meeting, 1993*. Groningen: University of Groningen.

Frank, L. K. (1925) 'Social problems', *American Journal of Sociology*, 30(4): 462–473.

Frank, L. K. (1936) 'Society as the patient', *American Journal of Sociology*, 42(3): 335–344.

Frank, L. K. (1939) 'Foreword', in E. V. Cowdry (ed.) *Problems of Ageing*, pp. xiii–xviii. Baltimore: Williams & Wilkins.

Frank, L. K. (1946) 'Gerontology', *Journal of Gerontology*, 1(1): 1–11.

French, J. R., Caplan, R. D. and Van Harrison, R. (1982) *The Mechanisms of Job Stress and Strain*. Chichester and New York: J. Wiley.

Fries, J. F. (1980) 'Aging, natural death, and the compression of morbidity', *New England Journal of Medicine*, 303(3): 130–35.

Fries, J. F. and Crapo, L. M. (1981) *Vitality and Aging: Implications of the Rectangular Curve*. San Francisco: W. H. Freeman.

Galton, Francis (1904) 'Eugenics: Its definition, scope, and aims', *American Journal of Sociology*, 10(1): 1–25.

Gillespie, R. (1987) 'Industrial fatigue and the discipline of physiology', in *Physiology in the American Context 1850–1940*, pp. 237–262. New York: Springer.

Giroux, E. (2008) 'Enquête de cohorte et analyse multivariée: Une analyse épisté-mologique et historique du rôle fondateur de l'étude de Framingham', *Revue d'Epidemiologie et de Santé Publique*, 56(3): 177–188.

Goldin, I., Cameron, G. and Balarajan, M. (2012) *Exceptional People: How Migration Shaped Our World and Will Define Our Future*. Princeton, NJ: Princeton University Press.

Gomart, E. and Hennion, A. (1999) 'A sociology of attachment: Music amateurs, drug users', *The Sociological Review*, 47(S1): 220–247.

Gompertz, B. (1825) 'On the nature of the function expressive of the law of human mor-tality, and on a new mode of determining the value of life contingencies', *Philosophical Transactions of the Royal Society of London*, 115(1825): 513–583.

Graham, H. (1991) 'The informal sector of welfare: A crisis in caring?' *Social Science & Medicine*, 32(4): 507–515.

Gruman, G. J. (2003) *A History of Ideas About the Prolongation of Life*. New York: Springer Publishing.

Habermas, J. (1984) *The Theory of Communicative Action, Vol. I*. Boston: Beacon.

Hacking, I. (1990) *The Taming of Chance*. Cambridge: Cambridge University Press.

Hannah, M. G. (2000) *Governmentality and the Mastery of Territory in Nineteenth-Century America*. Cambridge: Cambridge University Press.

Haraway, D. (1991) *Simians, Cyborgs, and Women: The Reinvention of Women*. London and New York: Routledge.

Haraway, D. (1997) *Modest_Witness@Second_Millennium. FemaleMan_Meets_OncoMouse: Feminism and Technoscience*. London and New York: Routledge.

Harper, S. (2006) *Ageing Societies*. London: Routledge.

Harris, J. (1997[1977]) *William Beveridge: A Biography*. Oxford: Oxford University Press.

Harvey, D. (2005) *A Brief History of Neoliberalism*. Oxford: Oxford University Press.

Hayden, E. (2015) 'Anti-ageing pill pushed as bona fide drug', *Nature*, 522(7556): 265–266.

Hayflick, L. (1998) 'How and why we age', *Experimental Gerontology*, 33(7): 639–653.

Haynes, S. G. and Feinleib, M. (1980a) 'Preface', in S. G. Haynes and M. Feinleib (eds) *Epidemiology of Aging*, pp. i–iii. Proceedings of the 2nd Conference (28–29 March 1977). Bethesda, MD: National Institutes of Health.

Haynes, S. G. and Feinleib, M. (eds) (1980b) *Epidemiology of Aging*. Proceedings of the 2nd Conference (28–29 March 1977). Bethesda, MD: National Institutes of Health.

Henderson, L. J. (1917) *The Order of Nature: An Essay*. Harvard, MA: Harvard University Press.

Hennock, E. P. (1987) *British Social Reform and German Precedents: The Case of Social Insurance, 1880–1914.* Oxford: Oxford University Press.

Heron, T. and Chown, S. (1967) *Age and Function.* London: Churchill.

Hirshbein, L. D. (2000) '"Normal" old age, senility, and the American Geriatrics Society in the 1940s', *Journal of the History of Medicine and Allied Sciences,* 55(4): 337–362.

Hoff, D. S. (2012) *The State and the Stork: The Population Debate and Policy Making in US History.* Chicago: University of Chicago Press.

Hofflander, A. E. (1966) 'The human life value: An historical perspective', *The Journal of Risk and Insurance,* 33(3): 381–391.

Hofman, A., Van Duijn, C. and Rocca, W. (1988) 'Is Alzheimer's disease distinct from normal ageing?' *The Lancet,* 332(8604): 226–227.

Holliday, R. (2007) *Aging: The Paradox of Life: Why We Age.* New York: Springer.

Hooker, P. F. (1965) 'Benjamin Gompertz 5 March 1779–14 July 1865', *Journal of the Institute of Actuaries,* 91(1965): 203–212.

House of Lords (2005) *Ageing: Scientific Aspects.* Available at: www.publications. parliament.uk/pa/ld200506/ldselect/ldsctech/20/20i.pdf (accessed 17 March 2009).

House of Lords (2006) *Science and Technology Committee: 6th Report.* Available at: www.publications.parliament.uk/pa/ld200506/ldselect/ldsctech/146/14603.htm (accessed 10 March 2009).

Huber, M., Knottnerus, J. A., Green, L., van der Horst, H., Jadad, A. R., Kromhout, D., Leonard, B., Lorig, K., Loureiro, M. I., van der Meer, J. W., Schnabel, P., Smith, R., van Weel, C. and Smid, H. (2011) 'How should we define health?' *British Medical Journal,* 343(2011): d4163.

Ilmarinen, J. (2006) 'The ageing workforce – challenges for occupational health', *Occupational Medicine,* 56(6): 362–364.

Ilmarinen, J. (2009) 'Work ability – a comprehensive concept for occupational health research and prevention', *Scandinavian Journal of Work, Environment & Health,* 35(1): 1–5.

Ilmarinen, J. (2010) *Sustainable Employability and Workability.* Maastricht: ESF-Age Network, pp. 18–19.

Ilmarinen, J., Tuomi, K., Eskelinen, L., Nygård, C. H., Huuhtanen, P. and Klockars, M. (1991) 'Background and objectives of the Finnish research project on aging workers in municipal occupations', *Scandinavian Journal of Work, Environment & Health,* 17(S1): 7–11.

Inglehart, J. K. (1978) 'The Carter Administration's health budget: Charting new priorities with limited dollars', *Milbank Memorial Fund Quarterly,* 56(1978): 51–57.

Ingram, D. K. (1988) 'Key questions in developing biomarkers of aging', *Experimental Gerontology,* 23(4–5): 429–434.

Ipsos Mori (2006) *Public Consultation on Ageing: Research into Public Attitudes Towards BBSRC and MRC-Funded Research on Ageing.* Available at: www.bbsrc.ac. uk/society/dialogue/attitude/ageing_mori_sri.pdf (accessed 27 March 2009).

Irwin, A. (2006) 'The politics of talk coming to terms with the "new" scientific governance', *Social Studies of Science,* 36(2): 299–320.

Irwin, A. and Horst, M. (2015) 'Engaging in a decentred world', in Chilvers and Kearnes (eds) *Remaking Participation: Science, Environment and Emergent Publics,* pp. 64–80. Abingdon: Routledge.

Jackson, M. (2012) 'The pursuit of happiness: The social and scientific origins of Hans Selye's natural philosophy of life', *History of the Human Sciences*, 25(5): 13–29.

Joyce, K. and Loe, M. (2010) 'A sociological approach to ageing, technology and health', *Sociology of Health & Illness*, 32(2): 171–180.

Joyce, K. and Mamo, L. (2006) *Graying the Cyborg. Age Matters: Realigning Feminist Thinking.* New York: Routledge.

Joyce, K., Loe, M. and Diamond-Brown, L. (2015) 'Science, technology and ageing', *Routledge Handbook of Cultural Gerontology*, pp. 157–63. London: Sage.

Kalache, A. and Gray, J. A. M. (1985) 'Health problems of older people in the developing world', in *Principles and Practice of Geriatric Medicine*, pp. 1279–1287. Chichester: John Wiley.

Kalache, A. and Kickbusch, I. (1997) 'A global strategy for healthy ageing', *World Health*, 50(4): 4–5.

Kalache, A., Barreto, S. M. and Keller, I. (2005) 'Global ageing: The demographic revolution in all cultures and societies', in M. Johnson (ed.) *The Cambridge Handbook of Age and Ageing*, pp. 30–47. Cambridge: Cambridge University Press.

Katz, S. (1996) *Disciplining Old Age: The Formation of Gerontological Knowledge.* Charlottesville: University Press of Virginia.

Katz, S. (2000) 'Busy bodies: Activity, aging, and the management of everyday life', *Journal of Aging Studies*, 14(2): 135–152.

Katz, S. and Marshall, B. (2003) 'New sex for old: Lifestyle, consumerism, and the ethics of aging well', *Journal of Aging Studies*, 17(1): 3–16.

Katz, S., Ford, A. B. and Moskovitz, R. W. (1963) 'Studies of illness in the aged: The index of Activities of Daily Living', *JAMA*, 185(Sept.): 914–919.

Katzman, R. (1976) 'The prevalence and malignancy of Alzheimer disease: A major killer', *Archives of Neurology*, 33(4): 217.

Katzman, R., Terry, R. D. and Bick, K. L. (eds) (1978) *Alzheimer's Disease: Senile Dementia and Related Disorders* (Vol. 7). Baltimore: Raven Press.

Kaufman, S. R. (1994) 'Old age, disease, and the discourse on risk: Geriatric assessment in US health care', *Medical Anthropology Quarterly*, 8(4): 430–447.

Kaufman, S. R., Shim J. K. and Russ, A. L. (2004) 'Revisiting the biomedicalization of aging: Clinical trends and ethical challenges', *Gerontologist*, 44(6): 731–738.

Kearns, R. A. and Andrews, G. J. (2004) 'Placing ageing: Positionings in the study of older people', in G. J. Andrews and D. R. Phillips (eds) *Ageing and Place: Perspectives, Policy, Practice*, pp. 13–23. London and New York: Routledge.

Keating, P. and Cambrosio, A. (2003) *Biomedical Platforms: Realigning the Normal and the Pathological in Late Twentieth-Century Medicine.* Cambridge, MA: MIT Press.

Kennedy, B. K., Berger, S. L., Brunet, A., Campisi, J., Cuervo, A. M., Epel, E. S., Franceschi, C., Lithgow, G. J., Morimoto, R. I., Pessin, J. E., Rando, T. A., Richardson, A., Schadt, E. E., Wyss-Coray, T. and Sierra, F. (2014) 'Geroscience: Linking aging to chronic disease', *Cell*, 159(4): 709–713.

Keynes, J. M. (1937) 'Some economic consequences of a declining population', *The Eugenics Review*, 29(1): 13–20.

Kingsland, S. (1982) 'The refractory model: The logistic curve and the history of population ecology', *Quarterly Review of Biology*, 57(1): 29–52.

Kirkwood, T. B. (1977) 'Evolution of ageing', *Nature*, 270(5635): 301–304.

Kirkwood, T. B. (1999) *Time of Our Lives: The Science of Human Aging.* Oxford: Oxford University Press.

Kirkwood, T. B. (2015) 'Deciphering death: A commentary on Gompertz' (1825) "On the nature of the function expressive of the law of human mortality, and on a new mode of determining the value of life contingencies"', *Phil. Trans. R. Soc. B*, 370(1666): 214–379.

Kirkwood, T. B. and Austad, S. N. (2000) 'Why do we age?', *Nature*, 408(6809): 233–238.

Koenig, B. A. (1988) 'The technological imperative in medical practice: The social creation of a "routine" treatment', in Lock and Gordon (eds) *Biomedicine Examined*, pp. 465–496. Dordrecht: Springer.

Kohli, M. (1986) 'The world we forgot: An historical review of the life course', in V. Marshall (ed.) *Later Life: The Social Psychology of Aging*. Beverly Hills, CA: Sage.

Kohli, M. (2007) 'The institutionalization of the life course: Looking back to look ahead', *Research in Human Development*, 4(3–4): 253–271.

Korenchevski, V. (1961) *Physiological and Pathological Aging*. Basel: Karger.

Lampland, M. and Star, S. L. (2009) *Standards and Their Stories: How Quantifying, Classifying, and Formalizing Practices Shape Everyday Life*. Ithaca, NY: Cornell University Press.

Laslett, P. (1989) *A Fresh Map of Life*. London: Weidenfeld & Nicolson,

Lassen, A. J. and Moreira, T. (2014) 'Unmaking old age: Political and cognitive formats of active ageing', *Journal of Aging Studies*, 30(1): 33–46.

Latour, B. (1987) *Science in Action*. Boston: Harvard University Press.

Latour, B. (1990) 'Drawing things together', in M. Lynch and S. Woolgar (eds) *Representation in Scientific Practice*. Cambridge, MA: MIT Press.

Latour, B. (1997) 'On actor-network theory: A few clarifications'. Available at: www.nettime.org/Lists-Archives/nettime-l-9801/msg00019.html (accessed 10 September 2001).

Latour, B. (2004) 'Why has critique run out of steam? From matters of fact to matters of concern', *Critical Inquiry*, 30(2): 225–248.

Latour, B. (2005) *Reassembling the Social – An Introduction to Actor-Network-Theory*. New York: Oxford University Press.

Law, J. (1994) *Organizing Modernity*. Oxford: Blackwell.

Law, J. (2002) *Aircraft Stories: Decentering the Object in Technoscience*. Durham, NC: Duke University Press.

Law, J. (2004) *After Method: Mess in Social Science Research*. London: Routledge.

Law, J. (2008) 'On sociology and STS', *The Sociological Review*, 56(4): 623–649.

Law, J. (2015) 'What's wrong with a one-world world?' *Distinktion: Scandinavian Journal of Social Theory*, 16(1): 126–139.

Law, J. and Mol, A. (1995) 'Notes on materiality and sociality', *The Sociological Review*, 43(2): 274–294.

Law, J. and Mol, A. (eds) (2002) *Complexities: Social Studies of Knowledge Practices. Science & Cultural Theory*. Durham, NC: Duke University Press.

Lawton, M. P. (1970) 'Assessment, integration, and environments for older people', *The Gerontologist*, 10(1 Part 1): 38–46.

Lawton, M. P. (1983) 'Environment and other determinants of well-being in older people', *The Gerontologist*, 23(4): 349–357.

Lawton, M. P. (1990) 'Aging and performance of home tasks', *Human Factors: The Journal of the Human Factors and Ergonomics Society*, 32(5): 527–536.

Lawton, M. P. and Brody, E. M. (1969) 'Assessment of older people: Self-maintaining and instrumental activities of daily living', *The Gerontologist*, 9(3): 179–186.

Lawton, M. P. and Simon, B. (1968) 'The ecology of social relationships in housing for the elderly', *Gerontologist*, 8(2): 108–115.

Lazarus, R. S. (1990) 'Theory-based stress measurement', *Psychological Inquiry*, 1(1): 3–13.

Lee, N. and Brown, S. (1994). Otherness and the actor network: The undiscovered continent', *The American Behavioral Scientist*, 37(6): 772.

Lenoir, R. (1979) 'L'invention du "troisième âge"', *Actes de la recherche en sciences sociales*, 26(1): 57–82.

Leonard, T. C. (2005) 'Retrospectives: Eugenics and economics in the Progressive Era', *The Journal of Economic Perspectives*, 19(4): 207–224.

Levy, D. M. and Peart, S. J. (2004) 'Statistical prejudice: From eugenics to immigrants', *European Journal of Political Economy*, 20(1): 5–22.

Lindsley, O. R. (1964) 'Geriatric behavioural prosthetics', in R. Kastenbaum (ed.) New *Thoughts on Old Age*. New York: Springer.

Lisbon Council (2000) Available at: www.euractiv.com/en/future-eu/lisbon-agenda/article-117510 (accessed 10 April 2015).

Loader, B. D., Hardey, M. and Keeble, L. (eds) (2009) *Digital Welfare for the Third Age: Health and Social Care Informatics For Older People*. London: Routledge.

Lock, M. (2013) *The Alzheimer Conundrum: Entanglements of Dementia and Aging*. Princeton, NJ: Princeton University Press.

Lockett, B. A. (1983) *Aging, Politics, and Research: Setting the Federal Agenda for Research on Aging*. New York: Springer.

Ludwig, F. C. (1980) 'What to expect from gerontological research?' *Science* (New York, NY), 209(4461): 1071.

Ludwig, F. C. and Smoke, M. E. (1980) 'The measurement of biological age', *Experimental Aging Research*, 6(6): 497–522.

Lyotard, J. F. (1979) *La condition postmoderne*. Paris: Les éditions de Minuit.

M'charek, A., Schramm, K. and Skinner, D. (2014) 'Topologies of race: Doing territory, population and identity in Europe', *Science, Technology & Human Values*, 39(4): 468–487.

McClearn, G. E. (1989) 'Biomarker characteristics and research on the genetics of aging', in D. E. Harrison (ed.) *Genetic Effects on Aging II*, pp. 234–256. Caldwell, NJ: Telford Press.

McCreadie, C. (2010) 'Technology and older people', *The Sage Handbook of Social Gerontology*, pp. 607–617. London: Sage.

McFarland, R. A. (1943) 'The older worker in industry', *Harvard Business Review*, 21(4): 34–35.

McFarland, R. A. (1946) *Human Factors in Air Transport Design*. New York: McGraw-Hill.

McFarland, R. A. (1956) 'The psychological aspects of aging', *Bulletin of the New York Academy of Medicine*, 32(1): 14–32.

McFarland, R. A. (1973) 'The need for functional age measurements in industrial gerontology', *Industrial Gerontology*, 19(1973): 1–19.

McFarland, R. A. and O'Doherty, B. M. (1959) 'Work and occupational skills', in James Birren (ed.) *Handbook of the Aging and the Individual*, pp. 452–500. Chicago: University of Chicago Press.

McFarland, R. A., Tune, G. S. and Welford, A. T. (1964) 'On the driving of automobiles by older people', *Journal of Gerontology*, 19(2): 190–197.

MacInnes, J. and Pérez Díaz, J. (2009) 'Transformations of the world's population', *The Routledge International Handbook of Globalization Studies*, p. 137. Abingdon, UK: Routledge.

MacKenzie, D. (1976) 'Eugenics in Britain', *Social Studies of Science*, 6(3/4): 499–532.

McKeown, T. (1979) *The Role of Medicine: Dream, Mirage or Nemesis?* (2nd edn). Oxford: Basil Blackwell.

McKhann, G., Drachman, D., Folstein, M., Katzman, R., Price, D. and Stadlan, E. M. (1984) 'Clinical diagnosis of Alzheimer's disease. Report of the NINCDS-ADRDA Work Group under the auspices of the Department of Health and Human Services Task Force on Alzheimer's Disease', *Neurology*, 34(7): 939–944.

Macnicol, J. (2015) *Neoliberalising Old Age*. Cambridge: Cambridge University Press.

Malthus, T. R. (1872/1792) *An essay on the principle of population; or, a view of its past and present effects on human happiness; with an inquiry into our prospects respecting the future removal or mitigation of the evils which it occasions*. London: Reeves and Turner.

Manton, K. G. (1982) 'Changing concepts of morbidity and mortality in the elderly population', *The Milbank Memorial Fund Quarterly*, 60(2): 183–244.

Manton, K. G. (1991) The dynamics of population aging: Demography and policy analysis', *The Milbank Memorial Fund Quarterly*, 69(2): 309–338.

Marks, H. (1997) *The Progress of Experiment: Science and Therapeutic Reform in the United States, 1900–1990*. Cambridge: Cambridge University Press.

Marmor, T. R. (1973) *The Politics of Medicare*. Chicago: Aldine Publishing.

Marres, N. (2007) 'The issues deserve more credit: Pragmatist contributions to the study of public involvement in controversy', *Social Studies of Science*, 37(5): 759–780.

Marshall, A. (1890) *Principles of Political Economy*. New York: Macmillan.

Marshall, T. H. (1953) 'Social selection in the welfare state', *The Eugenics Review*, 45(2): 81.

Martin, Moira (1995) 'Medical knowledge and medical practice: Geriatric medicine in the 1950s', *Social History of Medicine*, 8(3): 443–461.

Mayo, E. (1933) *The Human Problems of an Industrial Organization*. New York: Macmillan.

Mechanic, D. (1977) 'The growth of medical technology and bureaucracy: Implications for medical care', *The Milbank Memorial Fund Quarterly. Health and Society*, 55(1): 61–78.

Medawar, P. B. (1946) 'Old age and natural death', *Modern Quarterly*, 1(30): 56.

Medawar, P. B. (1952) *An Unsolved Problem of Biology*. London: University College London.

Medawar, P. and Medawar, J. S. (1977) *The Life Science*. London: Wildhood House.

Miller, R. A. (2001) 'Biomarkers of Aging', *Science of Aging Knowledge Environment*, 2001(1): pe2.

Miller, R. (2002) 'Extending life: Scientific prospects and political obstacles', *Milbank Memorial Fund Quarterly*, 80(1): 155–174.

Miller, R. A. (2009) '"Dividends" from research on aging – can biogerontologists, at long last, find something useful to do?' *The Journals of Gerontology Series A: Biological Sciences and Medical Sciences*, 64(2): 157–160.

Milligan, C. (2012) *There's No Place Like Home: Place and Care in an Ageing Society*. Farnham: Ashgate Publishing.

Mol, A. (2002) *The Body Multiple: Ontology in Medical Practice*. Durham, NC: Duke University Press.

Mol, A. (2008) *The Logic of Care: Health and the Problem of Patient Choice*. London and New York: Routledge.

Mol, A. (2012) 'Mind your plate! The ontonorms of Dutch dieting', *Social Studies of Science*, 43(3): 379–396.

Mol, A. and Law, J. (1994) 'Regions, networks and fluids: Anaemia and social topology', *Social Studies of Science*, 24(4): 641–671.

Moody, H. R. (2006) *Aging: Concepts and Controversies*. Thousand Oaks, CA: Pine Forge Press.

Morehouse, L. E. (1959) 'The strength of a man', *Human Factors: The Journal of the Human Factors and Ergonomics Society*, 1(2): 43–48.

Moreira, T. (2000) 'Translation, difference and ontological fluidity: Cerebral angiography and neurosurgical practice (1926–45)', *Social Studies of Science*, 30(3): 421–446.

Moreira, T. (2004) 'Surgical monads: A social topology of the operating room', *Environment and Planning D: Society and Space*, 22(1): 53–69.

Moreira, T. (2005) Diversity in clinical guidelines: The role of repertoires of evaluation', *Social Science & Medicine*, 60(9): 1975–1985.

Moreira, T. (2006) 'Heterogeneity and coordination of blood pressure in neurosurgery', *Social Studies of Science*, 36(1): 69–97.

Moreira, T. (2009) 'Hope and truth in drug development and evaluation in Alzheimer's Disease', in Jesse Ballenger, Peter Whitehouse, Constantine Lyketsos, Peter Rabins and Jason Karlawish (eds) *Treating Dementia: Do We Have a Pill for It?*, pp. 210–230. Baltimore: Johns Hopkins University Press.

Moreira, T. (2010) 'Ageing, Health and Justice', in Thomas Mathar and Yvonne J. F. M. Jansen (eds) *Health Promotion and Prevention Programmes in Practice*, pp. 5–32. Berlin/San Francisco: Transcript/Transaction.

Moreira, T. (2012a) *The Transformation of Contemporary Health Care: The Market, the Laboratory, and the Forum*. New York: Routledge.

Moreira, T. (2012b) 'Health care standards and the politics of singularities shifting in and out of context', *Science, Technology & Human Values*, 37(4): 307–331.

Moreira, T. and Bond, J. (2008) 'Does the prevention of brain ageing constitute anti-ageing medicine? Outline of a new space of representation for Alzheimer's Disease', *Journal of Aging Studies*, 22(4): 356–365.

Moreira, T. and Palladino, P. (2005) 'Between truth and hope: On Parkinson's disease, neurotransplantation and the production of the "self" ', *History of the Human Sciences*, 18(3): 55–82.

Moreira, Tiago and Palladino, Paolo (2008) 'Squaring the curve: The anatomo-politics of ageing, life and death', *Body & Society*, 14(3): 21–47.

Moreira, T. and Palladino, P. (2012) 'Questions of life and death: A genealogy', in B. S. Turner (ed.) *Routledge Handbook of Body Studies*. New York: Routledge.

Moreira, T., Hughes, J. C., Kirkwood, T., May, C., McKeith, I. and Bond, J. (2008) 'What explains variations in the clinical use of mild cognitive impairment (MCI) as a diagnostic category?' *International Psychogeriatrics*, 20(4): 697–709.

Moreira, T., May, C. and Bond, J. (2009) 'Regulatory objectivity in action: Mild cognitive impairment and the collective production of uncertainty', *Social Studies of Science*, 39(5): 665–690.

Moriyama, I. M. (1964) *The Change in Mortality Trends in the US. Vital and Health Statistics*. Washington, DC: National Center for Health Statistics.

Moriyama, I. M. (1968) 'Problems in the measurement of health status', in E. B. Sheldon and W. E. Moore (eds) *Indicators of Social Change*, pp. 573–600. New York: Russell Sage Foundation.

Morschhäuser, M. and Sochert, R. (2007) *Beschäftigungsfähigkeit erhalten. Strategien und Instrumente für ein langes gesundes Arbeitsleben. (Hg.)*. Essen: BKK Bundesverband.

Mort, M., Roberts, C. and Callén, B. (2013) 'Ageing with telecare: Care or coercion in austerity?' *Sociology of Health & Illness*, 35(6): 799–812.

Mort, M., Roberts, C., Pols, J., Domènech, M. and Moser, I. (2015) 'Ethical implications of home telecare for older people: A framework derived from a multisited participative study', *Health Expectations*, 18(3): 438–449.

Moser, Ingunn (2008) 'Making Alzheimer's disease matter: Enacting, interfering and doing politics of nature', *Geoforum*, 39(1): 98–110.

Moulaert, T. and Paris, M. (2013) 'Social policy on ageing: The case of "active ageing" as a theatrical metaphor', *International Journal of Social Science Studies*, 1(2): 113–123.

Moynihan, R., Heath, I. and Henry, D. (2002) 'Selling sickness: The pharmaceutical industry and disease mongering. Commentary: Medicalisation of risk factors', *British Medical Journal*, 324(7342): 886–891.

Mullan, P. (2000) *The Imaginary Time Bomb. Why an Ageing Population is Not a Social Problem*. London: IB Tauris.

Murray, I. M. (1951) 'Assessment of physiological age by combination of several criteria: Vision, hearing, blood pressure, and muscle force', *Journal of Gerontology*, 6(1): 120–126.

Mykytyn, C. E. (2006) 'Anti-aging medicine: A patient/practitioner movement to redefine aging', *Social Science & Medicine*, 62(3): 643–653.

Mykytyn, C. E. (2010) 'A history of the future: The emergence of contemporary anti-ageing medicine', *Sociology of Health & Illness*, 32(2): 181–196.

Myrdal, G. (1944) *An American Dilemma: The Negro Problem and American Democracy*. New York: Harper.

Nahemow, L. and Lawton, M. P. (1973) 'Toward an ecological theory of adaptation and aging', in W. Preiser (ed.) *Environmental Design Research*, pp. 24–32. Stroudsburg, PA: Dowden, Hutchinson & Ross.

Neilson, B. (2003) 'Globalization and the biopolitics of aging', *CR: The New Centennial Review*, 3(2): 161–186.

Neilson, B. (2009) 'Ageing and globalisation in a moment of so-called crisis', *Health Sociology Review*, 18(4): 349–363.

Neilson, B. (2012) 'Ageing, experience, biopolitics: Life's unfolding', *Body & Society*, 18(3–4): 44–71.

Nesse, R. M. and Williams, G. C. (1996) *Why We Get Sick: The New Science of Darwinian Medicine*. New York: Vintage.

Neugarten, B. L. (1964) *Personality in Middle and Late Life: Empirical Studies*. Oxford: Atherton.

Neugarten, B. L. (1968) *Middle Age and Aging: A Reader in Social Psychology*. Chicago, ILL: University of Chicago Press.

Neugarten, B. L. (1974) 'Age groups in American Society and the rise of the young-old', *The ANNALS of the American Academy of Political and Social Science*, 415(1): 187–198.

Neugarten, B. L. and Havighurst, R. J. (1976) *Social Policy, Social Ethics, and the Aging Society*. Chicago: Committee on Human Development, University of Chicago Press.

Neugarten, B. L., Moore, J. W. and Lowe, J. C. (1965) 'Age norms, age constraints, and adult socialization', *American Journal of Sociology*, 70(6): 710–717.

Nowotny, H. (1994) *Time: The Modern and Postmodern Challenge*. Cambridge: Polity Press.

Nowotny, H., Scott, P. and Gibbons, M. (2001) *Re-thinking Science: Knowledge Production in an Age of Uncertainty*. Cambridge: Polity.

O'Malley, P. (2002) 'Imagining insurance: Risk, thrift and life insurance in Britain', in T. Baker and J. Simon (eds) *The Changing Culture of Insurance and Responsibility*, pp. 97–115. Chicago: University of Chicago Press.

O'Malley, P. (2009) 'Uncertainty makes us free: Liberalism, risk and individual security', *Behemoth. A Journal on Civilisation*, 2(3): 24–38.

Oberlander, J. (2003) *The Political Life of Medicare*. Chicago: University of Chicago Press.

OECD (1994) *New Orientations for Social Policy: Social Policy Studies 12*. Paris: Organisation for Economic Co-operation and Development.

Olshansky, S. J. and Ault, A. B. (1986) 'The fourth stage of the epidemiologic transition: The age of delayed degenerative diseases', *The Milbank Quarterly*, 64 (3): 355–391.

Olshansky, S., Perry, D., Miller, R. A. and Butler, R. N. (2007) 'Pursuing the longevity dividend', *Annals of the New York Academy of Sciences*, 1114(1): 11–13.

Olshansky, J. S., Biggs, S., Achenbaum, A. W., Davison, G. C., Fried, L., Gutman, G., Kalache, A., Khaw, K. T., Fernandez, A., Rattan, S. I. S., Guimarães, R. M., Milner, C. and Butler, R. N. (2011) 'The global agenda council on the ageing society: Policy principles', *Global Policy*, 2(1): 97–105.

Olshansky, S. J., Beard, J. and Börsch-Supan, A. (2012) 'The longevity dividend: Health as an investment', in J. R. Beard, S. Biggs, D. E. Bloom, L. P. Fried, P. Hogan, A. Kalache and S. J. Olshansky (eds) *Global Population Ageing: Peril or Promise?*, pp. 57–60. Geneva, Switzerland: World Economic Forum.

Omran, A. R. (1971) 'The epidemiologic transition: A theory of the epidemiology of population change', *The Milbank Memorial Fund Quarterly*, 49(4): 509–538.

Oppenheimer, Gerald M. (2006) 'Profiling risk: The emergence of coronary heart disease epidemiology in the United States (1947–70)', *International Journal of Epidemiology*, 35(3): 720–730.

Osborne, T. and Rose, N. (2008) 'Populating sociology: Carr-Saunders and the problem of population', *The Sociological Review*, 56(4): 552–578.

Östlund, B., Olander, E., Jonsson, O. and Frennert, S. (2015) 'STS-inspired design to meet the challenges of modern aging. Welfare technology as a tool to promote user driven innovations or another way to keep older users hostage?' *Technological Forecasting and Social Change*, 93(2015): 82–90.

Parascandola, J. (1971) 'Organismic and holistic concepts in the thought of LJ Henderson', *Journal of the History of Biology*, 4(1): 63–113.

Parascandola, M. (2004) 'Skepticism, statistical methods, and the cigarette: A historical analysis of a methodological debate', *Perspectives in Biology and Medicine*, 47(42): 244–261.

Park, H. W. (2008) 'Edmund Vincent Cowdry and the making of gerontology as a multidisciplinary scientific field in the United States', *Journal of the History of Biology*, 41(3): 529–572.

Park, H. W. (2016) *Old Age, New Science: Gerontologists and Their Biosocial Visions*. Pittsburgh: University of Pittsburgh Press.

Park, R. E. (1928) 'Human migration and the marginal man', *American Journal of Sociology*, 33(6): 881–893.

Partridge, L. (2010) 'The new biology of ageing', *Philosophical Transactions of the Royal Society of London B: Biological Sciences*, 365(1537): 147–154.

Patnoe, S. (2013) *A Narrative History of Experimental Social Psychology: The Lewin Tradition*. New York: Springer.

Paul, D. (1995) *Controlling Human Heredity, 1865 to the Present*. Amherst: Humanity Press.

Paul, D. (2001) 'History of Eugenics', in Neil Smelser and Paul Baltes (eds) *International Encyclopedia of Social and Behavioral Sciences*, pp. 4896–4901. Amsterdam: Elsevier.

Pearson, K. (1909) *The Groundwork of Eugenics, Eugenics Laboratory Lecture Series 2.* London: Dular and Company.

Peine, A., Faulkner, A., Jaeger, B. and Moors, E. (2015) 'Science, technology and the "grand challenge" of ageing – understanding the socio-material constitution of later life', *Technological Forecasting and Social Change*, 93(April): 1–9.

Perry, E. K., Tomlinson, B. E., Blessed, G., Bergmann, K., Gibson, P. H. and Perry, R. H. (1978) 'Correlation of cholinergic abnormalities with senile plaques and mental test scores in senile dementia', *British Medical Journal*, 2(6150): 1457–1459.

Picard, R. and Mouchet, S. (2009) *La métamorphose de la médecine*. Paris: PUF.

Piekkola, H. (2004) *Active Ageing Policies in Finland* (No. 898), ETLA Discussion Papers. Helsinki: The Research Institute of the Finnish Economy (ETLA).

Pollack, A. (2011) 'A Blood Test Offers Clues to Longevity', *New York Times*, 12 May: B1.

Pols, J. and Willems, D. (2011) 'Innovation and evaluation: Taming and unleashing telecare technology', *Sociology of Health & Illness*, 33(3): 484–498.

Porter, D. (1997) 'The decline of social medicine in Britain in the 1960s', in Porter, Dorothy (ed.) *Social Medicine and Medical Sociology in the Twentieth Century*, pp. 97–119. Amsterdam: Rodopi.

Prescott, H. M. (2004) '"I was a teenage dwarf": The social construction of "normal" adolescent growth and development in the United States', in A. M. Stern and H. Markel (eds) *Formative Years: Children's Health in the United States, 1880–2000*, pp. 153–184. Ann Arbor: University of Michigan Press.

Rabinach, A. (1990) *The Human Motor: Energy, Fatigue, and the Origins of Modernity*. New York: Basic Books.

Rabinow, P. (2000) 'Epochs, presents, events' in A. Cambrosio, M. Lock and A. Young (eds) *Living and Working with the New Medical Technologies: Intersections of Inquiry*, pp. 8–31. Cambridge: Cambridge University Press.

Rabinow, P. (2005) *Artificiality and Enlightenment: From Sociobiology to Biosociality*. Oxford: Blackwell.

Rabinow, P. and Rose, N. (2006) 'Biopower today', *BioSocieties*, 1(2): 195–217.

Ramsden, E. (2002) 'Carving up population science: Eugenics, demography and the controversy over the "biological law" of population growth', *Social Studies of Science*, 32(5–6): 857–899.

Rattan, S. I. (2002) A global view of the causes of ageing: An interview with Robin Holliday. *Biogerontology*, 3(5): 317–324.

Renwick, C. (2012) *British Sociology's Lost Biological Roots: A History of Futures Past*. Basingstoke: Palgrave Macmillan.

Rip, A. (2002) 'Regional innovation systems and the advent of strategic science', *The Journal of Technology Transfer*, 27(1): 123–131.

Robertson, A. (1990) 'The politics of Alzheimer's disease: A case study in apocalyptic demography', *International Journal of Health Services*, 20(3): 429–442.

Rogers, W. A. and Fisk, A. D. (2010) 'Toward a psychological science of advanced technology design for older adults', *The Journals of Gerontology Series B: Psychological Sciences and Social Sciences*, 65(6): 645–653.

Rose, N. (2009) *The Politics of Life Itself: Biomedicine, Power, and Subjectivity in the Twenty-First Century*. Princeton, NJ: Princeton University Press.

Rose, N. and Novas, C. (2005) 'Biological citizenship', in A. Ong and S. Collier (eds) *Global Assemblages: Technology, Politics and Ethics as Anthropological Problems*, pp. 439–463. Malden, MA: Blackwell.

Ross, A. and Lloyd, J. (2012) *Who Uses Telecare?* London: The Strategic Society Centre.

Rosset, E. (1964) *Aging Process of Populations*. New York: Macmillan.

Roth, M., Tomlinson, B. E. and Blessed, G. (1966) 'Correlation between scores for dementia and counts of "senile plaques" in cerebral grey matter of elderly subjects', *Nature*, 209(5018): 109.

Rothstein, W. G. (2003) *Public Health and the Risk Factor: A History of an Uneven Medical Revolution*. Rochester: University of Rochester Press.

Rowe, J. W. and Kahn, R. L. (1987) 'Human aging: Usual and successful', *Science*, 237(4811): 143–149.

Rozanova, J. (2010) 'Discourse of successful aging in The Globe & Mail: Insights from critical gerontology', *Journal of Aging Studies*, 24(4): 213–222.

Rutherford, A. (2009) *Beyond the Box: B. F. Skinner's Technology of Behaviour from Laboratory to Life, 1950s–1970s*. Toronto: University of Toronto Press.

Salthouse, T. A. (1986) 'Functional age', in J. Birren, P. Robinson and J. Livingston (eds) *Age, Health, and Employment*, pp. 78–92. New York: Prentice Hall.

Sauvy, A. (1950) 'Besoins et possibilités de l'immigration française (1re partie)', *Population* (French edition), 5(3): 209–228.

Sauvy, A. (1961) *Le 'tiers-monde': sous-développement et développement*. Paris: Presses universitaires de France.

Schaie, K. W. (1973) 'Methodological problems in descriptive developmental research in adulthood and aging', in J. R. Nesselroade and H. W. Reese (eds) *Proceedings of the Conference on Life-span Developmental Psychology: Methodological Issues*, pp. 253–280. Morgantown: University of West Virginia.

Schaie, K. W. (1977) 'Functional age and retirement'. Paper presented at the Annual Meeting of the International Society for the Study of Behavioral Development (4th, Pavia, ITALY, 19–25 September 1977).

Schaie, K. W. and Baltes, P. B. (1975) 'On sequential strategies in developmental research', *Human Development*, 18(5): 384–390.

Scheffler, R. W. (2011) 'The fate of a progressive science: The Harvard Fatigue Laboratory, athletes, the science of work and the politics of reform', *Endeavour*, 35(2): 48–54.

Scheffler, R. W. (2015) 'The power of exercise and the exercise of power: The Harvard Fatigue Laboratory, distance running, and the disappearance of work, 1919–1947', *Journal of the History of Biology*, 48(3): 391–423.

Schneider, E. L. and Brody, J. A. (1983) 'Aging, natural death, and the compression of morbidity: Another view', *New England Journal of Medicine*, 309(14): 854–856.

Schumpeter, J. A. (1950) *Capitalism, Socialism, and Democracy, 3rd Ed.* [1962]. New York: Harper.

Scott, J. C. (1998) *Seeing Like a State: How Certain Schemes to Improve the Human Condition Have Failed*. Newhaven: Yale University Press.

Sealander, J. (1997) *Private Wealth and Public Life: Foundation Philanthropy and the Reshaping of American Social Policy from the Progressive Era to the New Deal*. Baltimore: Johns Hopkins University Press.

Semmel, B. (1958) 'Karl Pearson: Socialist and Darwinist', *The British Journal of Sociology*, 9(2): 111–125.

Serow, W. J. and Sly, D. F. (1988) 'The demography of current and future aging cohorts', in Committee on an Aging Population (ed.) *America's Aging: The Social and Built Environment in the Older Society*, pp. 42–102. Washington, DC: National Academy Press.

Settersten, R. A. and Mayer, K. U. (1997) 'The measurement of age, age structuring, and the life course', *Annual Review of Sociology*, 23(1): 233–261.

Shapin, S. (2008) *The Scientific Life: A Moral History of a Late Modern Vocation*. Chicago, ILL: University of Chicago Press.

Shils, E. (1957) 'Primordial, personal, sacred and civil ties: Some particular observations on the relationships of sociological research and theory', *The British Journal of Sociology*, 8(2): 130–145.

Shock, N. W. (1943) 'The effect of menarche on basal physiological functions in girls', *American Journal of Physiology*, 139(2): 288–292.

Shock, N. W. (1947) 'The United States Public Health Service-Baltimore City Hospitals Research Section on Gerontology', *Journal of Gerontology*, 2(2): 169–170.

Shock, N. W. (1956) 'Some physiological aspects of aging in man', *Bulletin of the New York Academy of Medicine*, 32(4): 268–283.

Shock, N. W. (1961) 'Public health and the aging population', *Public Health Reports*, 76(11): 1023–1027.

Shock, N. W., Greulich, R. C., Aremberg, D., Costa, P. T., Lakatta, E. G. and Tobin, J. D. (1984) *Normal Human Aging: The Baltimore Longitudinal Study of Aging*. Bethesda: National Institute of Aging.

Smith, J. M. (1958) 'Prolongation of the life of Drosophila subobscura by a brief exposure of adults to a high temperature', *Nature*, 181(4607): 496–497.

Smith, R. (1992) *Inhibition: History and Meaning in the Sciences of Mind and Brain*. Los Angeles: University of California Press.

Smuts, A. B. (2008) *Science in the Service of Children, 1893–1935*. New Haven: Yale University Press.

Sontag, L. W. (1971) 'The history of longitudinal research: Implications for the future', *Child Development*, 42(4): 987–1002.

Spektorowski, A. and Ireni-Saban, L. (2014) *Politics of Eugenics: Productionism, Population, and National Welfare*. Abingdon, Oxon: Routledge.

Sprott, Richard L. (1999) 'Biomarkers of aging', *Journals of Gerontology: Biological Sciences*, 54A(11): B464–B465.

Star, S. L. (1991) 'Power, technologies and the phenomenology of conventions: On being allergic to onions', in J. Law (ed.) *A Sociology of Monsters: Essays on Power. Technology and Domination*. London: Routledge.

Star, S. L. and Griesemer, J. (1989) 'Institutional ecology, "translations" and boundary objects: Amateurs and professionals in Berkley's Museum of Vertebrate Zoology 1907–1939', *Social Studies of Science*, 19(3): 387–420.

Stengers, I. (2005) 'The cosmopolitical proposal', in Latour and Weibel (eds) *Making Things Public: Atmospheres of Democracy*, pp. 994–998. Cambridge, MA: MIT Press.

Stern, N. (1978) 'Age and achievement in mathematics: A case-study in the sociology of science', *Social Studies of Science*, 8(1): 127–140.

Stewart, J. (2008) 'The political economy of the British National Health Service, 1945–1975: Opportunities and constraints?', *Medical History*, 52(4): 453–470.

Stone, Jane Livermore and Norris, Arthur H. (1966) 'Activities and attitudes of participants in the Baltimore Longitudinal Study', *Journal of Gerontology*, 21(4): 575–580.

Strathern, M. (2005[1991]) *Partial Connections*. London: Rowman Altamira.

Sullivan, D. F. (1966) 'Conceptual problems in developing an index of health', *USPHS Publ. No. 1000*, 2(17).

Szreter, S. (1993) 'The idea of demographic transition and the study of fertility change: A critical intellectual history', *Population and Development Review*, 19(4): 659–701.

Tellmann, U. (2013) 'Catastrophic populations and the fear of the future: Malthus and the genealogy of liberal economy', *Theory, Culture & Society*, 30(2): 135–155.

Thane, P. (1989) 'History and the sociology of ageing', *Social History of Medicine*, 2(1): 93–96.

Thane, P. (2000) *Old Age in English history: Past Experiences, Present Issues*. Oxford: Oxford University Press.

Thévenot, L. (1984) 'Rules and implements: Investment in form', *Social Science Information*, 23(1): 1–45.

Thévenot, L. (1990) 'La politique des statistiques: les origines sociales des enquêtes de mobilité sociale', *Annales E.S.C.*, 45(6): 1275–1300.

Thévenot, L. (2006) 'Convention school', in J. Beckert and M. Zafirovski (eds) *International Encyclopedia of Economic Sociology*, pp. 111–115. London: Routledge.

Thévenot, L. (2007) 'The plurality of cognitive formats and engagements moving between the familiar and the public', *European Journal of Social Theory*, 10(3): 409–423.

Thévenot, L. (2009) Postscript to the special issue: Governing life by standards: A view from engagements, *Social Studies of Science*, 39(5): 793–813.

Thompson, W. S. (1929) 'Recent trends in world population', *American Journal of Sociology*, 34(6): 959–979.

Tilly, C. (1978) 'Migration in modern European history', in W. H. McNeill and B. Adams (eds) *Human Migration: Patterns and Policies*, pp. 48–72. Bloomington, IN: Indiana University Press.

Townsend, P. (1957) *The Family Life of Old People: An Inquiry in East London*. London: Routledge.

Townsend, P. (1962) *The Last Refuge: A Survey of Residential Institutions and Homes for the Aged in England and Wales*. London: Routledge.

Townsend, P. (1981) 'The structured dependency of the elderly: A creation of social policy in the twentieth century', *Ageing and Society*, 1(1): 5–28.

Townsend, P. (2009) 'A society for people', *Social Policy and Society*, 8(2): 147–158.

Treas, J. (2009) 'Age in standards and standards for age: Institutionalizing chronological age as biographical necessity', in M. Lampland and S. L. Star (eds), *Standards and Their Stories: How Quantifying, Classifying, and Formalizing Practices Shape Everyday Life*. Cornell, NY: Cornell University Press.

Tulle, E. and Mooney, E. (2002) Moving to 'age-appropriate' housing: Government and self in later life', *Sociology*, 36(3): 685–702.

Tunbridge, R. (1955) 'Chairman's opening remarks', in G. Wolstenholme and M. Cameron (eds) *Ciba Foundation Colloquia on Ageing. General Aspects (I)*, pp. 1–3. Boston: Little, Brown.

Tuomi, K., Ilmarinen, J., Klockars, M., Nygård, C. H., Seitsamo, J., Huuhtanen, P., Martikainen, R. and Aalto, L. (1997) 'Finnish research project on aging workers in 1981–1992', *Scandinavian Journal of Work, Environment & Health*, 23(S1): 7–11.

Tuomi, K., Ilmarinen, J., Jahkola, A., Katajarinne, L. and Tulkki, A. (1998) *Work Ability Index (WAI)*. Helsinki: Finnish Institute of Occupational Health.

Turner, B. S. (1989) 'Ageing, status politics and sociological theory', *British Journal of Sociology*, 40(4): 588–606.

Turner, B. S. (2009) *Can We Live Forever?: A Sociological and Moral Inquiry*. London: Anthem Press.

United Nations (2001) *World Population Prospects: The 2000 Revision*. New York: United Nations.

UNPD (United Nations Population Division) (2001) *Replacement Migration: Is it a Solution to Declining and Ageing Populations?* New York: United Nations.

UNPF/HAI (United Nations Population Fund and HelpAge International) (2012) *Ageing in the Twenty-First Century: A Celebration and a Challenge*. New York and London: United Nations UNPF and HAI.

UNPoA/IAGG (United Nations Programme on Ageing/International Association of Gerontology and Geriatrics) (2007) *Research Agenda on Ageing for The Twenty-First Century*. New York: United Nations UNPoA and IAGG.

Van Houtum, H. (2010) 'Human blacklisting: The global apartheid of the EU's external border regime', *Environment and Planning D: Society and Space*, 28(6): 957–976.

Venturini, T. (2010) 'Diving in magma: How to explore controversies with actor-network theory', *Public Understanding of Science*, 19(3): 258–273.

Viner, R. (1999) 'Putting stress in life: Hans Selye and the making of stress theory', *Social Studies of Science*, 29(3): 391–410.

Wahl, H. W., Iwarsson, S. and Oswald, F. (2012) 'Aging well and the environment: Toward an integrative model and research agenda for the future', *The Gerontologist*, 52(3): 306–316.

Wahlberg, A. and Rose, N. (2015) 'The governmentalization of living: Calculating global health', *Economy and Society*, 44(1): 60–90.

Walker, A. (2002) 'A strategy for active ageing', *International Social Security Review*, 55(1): 121–139.

Walker, A. (ed.) (2014) *The New Science of Ageing*. Bristol: Policy Press.

Walker, F. A. (1873) 'Our population in 1900', *Atlantic Monthly*, 32(192): 487–495.

Walker, F. A. (1874) *Statistical Atlas of the United States Based on the Results of the 9th Census 1870*. New York: Julius Bien.

Walker, F. A. (1891) 'Immigration and Degradation', *Forum*, 11(August): 634–644.

Walker, F. A. (1896) 'Restriction of Immigration', *Atlantic Monthly*, 77(464): 822–829.

Walker, F. A. (1899) *Discussions in Economics and Statistics*, Vol. II. New York: Henry Holt and Co.

Wallerstein, I. (1996) *Open the Social Sciences: Report of the Gulbenkian Commission on the Restructuring of the Social Sciences*. Stanford: Stanford University Press.

Walters, W. (2002) 'Mapping Schengenland: Denaturalizing the border', *Environment and Planning D: Society and Space*, 20(5): 561–580.

Weber, M. (1978) *Economy and Society: An Outline of Interpretive Sociology*. Los Angeles: University of California Press.

Weisz, G. and Olszynko-Gryn, J. (2010) 'The theory of epidemiologic transition: The origins of a citation classic', *Journal of the History of Medicine and Allied Sciences*, 65(3): 287–326.

Welford, A. T. (1953) 'Extending the employment of older people', *British Medical Journal*, 2(4847): 1193.

Welford, A. T. (1958) *Ageing and Human Skill*. Oxford: Oxford University Press.

Welford, A. T. (1965) 'Performance, biological mechanisms and age: A theoretical sketch', in A. T. Welford and J. E. Birren (eds) *Behavior, Aging, and the Nervous*. Springfield, ILL: Thomas.

Welford, A. T. (1966) 'Industrial work suitable for older people: Some British studies', *The Gerontologist*, 6(1): 4–9.

Weyman, A., Wainwright, D., O'Hara, R., Jones, P. and Buckingham, A. (2012) *Extending Working Life: Behaviour Change Interventions*. London: Department for Work and Pensions.

Whitehouse, P. J., Price, D. L., Struble, R. G., Clark, A. W., Coyle, J. T. and Delon, M. R. (1982) 'Alzheimer's disease and senile dementia: Loss of neurons in the basal forebrain', *Science*, 215(4537): 1237–1239.

WHO (World Health Organization) (1993) 'Aging and work capacity. Report of a WHO Study Group', *WHO Technical Report Series 835*. Geneva: World Health Organization.

Wilkie, A. and Michael, M. (2009) 'Expectation and mobilisation enacting future users', *Science, Technology & Human Values*, 34(4): 502–522.

Willmott, P. and Young, M. D. (1957) *Family and Kinship in East London*. London: Routledge & Kegan Paul.

Wilson, D. (2014) 'Quantifying the quiet epidemic: Diagnosing dementia in late 20th-century Britain', *History of the Human Sciences*, 27(5): 126–146.

Wolfe, A. and Lausen, J. (1997) 'Identity Politics and the Welfare State', *Social Philosophy and Policy*, 14(2): 213–255.

Wolfenden, J. (1978) *The Future of Voluntary Organisations: Report of the Wolfenden Committee*. London: Croom Helm.

Wolstenholme, G. E. W. and Cameron, Margaret P. (eds) (1955) *Ciba Foundation Colloquia on Ageing: Volume 1: General Aspects*. Boston: Little, Brown.

Wolstenholme, G. E. W. and O'Connor, M. (eds) (1957) *Ciba Foundation Colloquia on Ageing: Volume 3: Methodological aspects*. Boston: Little, Brown.

World Bank (1994) *Averting the Old Age Crisis: Policies to Protect the Old and Promote Growth*. New York: Oxford University Press.

Wynne, B. and Felt, U. (2007) *Taking European Knowledge Society Seriously, Report of the Expert Group on Science and Governance to the Science, Economy and Society Directorate*. Brussels: European Commission.

Yeomans, L. (2011) *An Update of the Literature on Age and Employment*. London: Health and Safety Executive.

Zola, I. K. (1972) 'Medicine as an institution of social control', *The Sociological Review*, 20(4): 487–504.

Zola, I. K. (1988) 'Policies and programs concerning aging and disability: Toward a unifying agenda', in *The Economics and Ethics of Long-Term Care and Disability* (pp. 90–130). Washington, DC: American Enterprise Institute for Public Policy Research.

Zuiderent-Jerak, T. (2015) *Situated Intervention: Sociological Experiments in Health Care*. Cambridge, MA: MIT Press.

Archival Sources

Alex Comfort Papers in the University College London Library/National Archives: References to this resource will be coded as CMFT

Institut national de la santé et de la recherche médicale (INSERM) Archive: References to this resource will be coded INSERM

Nathan W. Shock Collection in the Bentley Historical Library at the University of Michigan. References to this resource will be coded as NWS.

Index

1000 Aviator Study 129